Champion Trees of Washington State

The Mineral Tree – the largest **Douglas-fir** tree ever recorded and the largest tree known to have grown in the state of Washington. Over 15 feet in diameter and 393' tall, it contained approximately 25,000 cubic feet of wood. Aged at 1,020 years, the Mineral Tree fell in a severe gale in 1930. A section of this tree may be seen at the Wind River Arboretum near Carson.

Champion Trees of Washington State

Robert Van Pelt

Washington State Big Tree Program
in association with
University of Washington Press

Western Hemlock, *Tsuga heterophylla*, the state tree of Washington.
This tree, on the upper Wynoochee trail in Olympic National Park, is the largest
known hemlock in the world.

Library of Congress Cataloging-in-Publication Data
Van Pelt, Robert.
    Champion trees of Washington State / Robert Van Pelt ; photos
and drawings by the author.
        p.   cm.
    Includes index.
    ISBN 0-295-97563-6 (pbk. : alk. paper)
    1. Trees—Washington (State)   I. Title.
SD383.3.U6V35   1996
582.1609797—dc20   96-13966
                                                                     CIP

The paper used in this publication is acid-free and recycled from 10
percent post-consumer and at least 50 percent pre-consumer waste. It
meets the minimum requirements of American National Standard for
Information Sciences—Permanence of Paper for Printed Library
Materials, ANSI Z39.48-1984. ∞

CONTENTS

Preface   vi

Acknowledgments   vii

Why Measure Trees?   viii

How to Measure a Tree   ix

Register of Washington State Big Trees

Native Trees   1

Introduced Trees   25

Washington Park Arboretum Map   102

Former Record Trees   103

Errata from 1994 Edition   105

Place Name Index   107

General Index   109

Big Tree Nomination Form   120

The **Pacific Madrona**, one of few native broadleaved evergreen trees, is famous for its red-orange bark which brightens up coastal areas throughout the Puget Sound region. The specimen shown, in downtown Port Angeles, is perhaps our handsomest. It is certainly among our largest.

PREFACE

Welcome to the world of big trees!

This is the fourth edition of the Washington State Big Tree Program list of champions. The program is now nine years old and includes an impressive list of many truly spectacular trees.

Since the last edition many new specimens have been added, both new species and larger trees of previously listed species. The list began with thirteen National Champion trees in 1987. The current edition contains some 1,350 individual trees belonging to 869 species or cultivated varieties. There were some deaths too, however. These and other former champions are listed on page 103. Trees from Washington that are the largest known examples of their kind in the country are also listed with the National Big Tree Program sponsored by the American Forestry Association (AFA, now called American Forests). The AFA has been keeping records on giant trees since 1940. National Record trees in Washington now number 45 and are denoted by a ❖ next to the AFA point total.

This book includes registers for native trees and introduced trees. The general index includes both common and scientific names, as well as general headings. The list of native trees is intended to be comprehensive, with nomenclature largely following *Checklist of United States Trees* by Elbert Little (1979). Introduced species, however, are so numerous that only a selection has been included. Nevertheless, the book now includes over 740 species and cultivars of introduced trees! Nomenclature of the myriad introduced trees listed follows *North American Landscape Trees* by Arthur Jacobson (1996). Recently published, this is the only work comprehensive enough to include nearly all of the introduced trees in this book. When the identification of a species or cultivar is in question, a (?) precedes the scientific name listed.

Whenever possible, precise locations (e.g., street addresses) are given. Directions to trees in city parks, national parks, and national forests are too complex to describe in the space provided; however, directions to any tree will be provided upon request. Write to: Washington Big Tree Program, Box 352100, Anderson Hall, University of Washington, Seattle, WA 98195. The whereabouts of trees in the Washington Park Arboretum in Seattle are described using the Arboretum's own grid notation system. A map of the Arboretum with grid is shown on page 102.

Kathy Van Pelt, Arthur Lee Jacobson, Ron Brightman, and I have perfected a technique we jokingly call 55 mph dendrology (identifying trees at a glance from a moving car), enabling us to explore large areas with minimal effort. Even so, large areas of the state remain unexplored, and it is only through the eyes of many that we learn the full extent of the rich heritage of trees our beautiful state has to offer. To find a champion tree, one first needs to know what the existing records are. The trees are all out there somewhere. Let's all get out and track them down.

Good luck, and good hunting!

Robert Van Pelt
Washington State Big Tree Program Coordinator
January 1996

ACKNOWLEDGMENTS —————————————

This book is dedicated to the late Randy Stoltmann, a fellow tree measurer who started the Big Tree Program for British Columbia. An enthusiastic tree hunter, Randy located many of the largest trees in Canada, as well as several record trees in Washington. He was also a great friend whose untimely death left a large vacancy in the British Columbia conservation community. His enthusiasm for tree exploration and conservation will be sorely missed.

In recognition of the people who participate in the program, I include their names beside the listed trees they have nominated. When I began the program in 1987, the list consisted of thirteen trees that were National Champions. Since then I have traveled throughout the state locating new and interesting trees to put on the list. Along the way I have met many interesting people, often with exciting stories to tell about their trees. Several people have assisted me in this project. Without them many of the trees listed might still be unknown.

The American Forestry Association sparked the idea of a big tree program when 56 years ago it started the National Big Tree Register. Washington residents occasionally contributed to this national program over the years. Up until 1987 there was no state coordinator for the national program, nor a state program. The College of Forest Resources at the University of Washington, particularly Drs. Jim Agee and Grant Sharpe, deserve credit for constant support which helped me to make this effort a reality.

Shirley Muse of Walla Walla has a great love of the trees in her area and started a heritage tree program there. She was also instrumental in producing a walking tour booklet of the record-sized trees in Walla Walla. She has nominated several trees and we have found many more together, many of which are listed in this publication.

Arthur Lee Jacobson, author of *Trees of Seattle* and *North American Landscape Trees* has taught me a great deal about tree identification since we met in 1986. His research on and searches for trees in Seattle led to the great abundance of Seattle trees on this list. His nominations appear frequently on the pages that follow, and are denoted by the initials ALJ.

The initials RGB also appear next to dozens of entries. Ron Brightman became known to me shortly before the second edition was released and nominated several trees for that book. His knowledge of trees is vast and is equalled only by his zeal for discovering new trees, which has resulted in greatly improved and expanded 1994 and 1996 editions.

Thanks also goes to Katherine Van Pelt, my best friend for several years. We met through Arthur Jacobson, having a shared love of trees in general as well as a special interest in the trees in and around Tacoma. Together we have explored the greater Tacoma area for noteworthy examples to put in a book we are writing on the subject. The initials KVP indicate trees she has discovered or nominated. Our kids appear in many of the photos, particularly Iris and Ivan.

Lastly, thanks goes to the many people who have nominated trees over the years. The publicity of the first three editions has led to many new nominations. While not all of them turned out to be new champions, many did, and it is always a pleasure to meet new people excited about trees.

WHY MEASURE TREES?

I have been asked many times "Why measure trees? Of what use is the information?" My usual response is "To see how big it is." I then go on to explain several uses for the information.

Big trees represent the maximum growth of a species. I feel more can be learned about a species by studying the extremes than the average. To understand the ecology of a species, for example, studying it at the extremes of its range provides a richer understanding of its limitations and requirements, helping to understand its behavior throughout the rest of its range.

From a scientific standpoint, recording the sizes of trees provides a basis for comparing growing conditions throughout the world. Many people moving here from other parts of the country don't realize that the growing conditions are ideal for many kinds of trees and may underestimate the potential of species they may have brought with them.

Tree measurements are also a useful source of information for landscape architects and homeowners who want to know how big a specific kind of tree can get, or want help in deciding where to plant it. Too often I see a tree planted too close to a house or sidewalk. Perhaps the tree looked good in that spot when first planted, but it can soon grow into unimagined proportions.

I became interested in measuring trees because I love them. Seeing a fully grown specimen thriving in the forest and knowing that it is the largest one of its kind is an unmatched experience.

Arthur Jacobson measuring the National Champion **Alpine Fir**. He is measuring it at 4½ feet above average ground level – in this case 3'4" from the high side (his position), and 5'8" from the low side.

HOW TO MEASURE A TREE⎯⎯⎯⎯

I have adopted internationally accepted standards for recording the size of a tree. These consist of a trunk measurement (circumference) and measurements of height and crown spread. A further measure, AFA (American Forestry Association) points, is based on the first three measurements.

Circumference

The first measurement is the simplest and the one most people think of when contemplating a tree's size. Ideally, a tree should be measured at breast height, which is 4 ½ feet or 1.37 meters above average ground level (Fig. 1).

Problems arise with the phrase 'above average ground level.' Many native trees grow on slopes or uneven ground; here the rule is to measure 4 ½ feet from both the high side and the low side, find the midpoint between the two, and take the measurement there (Fig. 2; photo, page viii). If the slope is so steep that the average location is below the ground level on the high side, then the lowest practical point is used (Fig. 3). Further complications arise when branches obstruct or influence the appropriate location. A tree with low branches (for example, an open-grown or city tree) might give a larger reading at 4 ½ feet than at 3 feet. The rule that the tree be measured at 4 ½ feet *or* the smallest reading obtainable below that point ensures that the smallest reading possible is used (Fig. 4; photo, page 73). Sprouting trees, such as hazels, frequently have several stems at 4 ½ feet. In such cases the largest stem is measured (Fig. 5).

Figure 1
Measuring a tree on flat ground

Figure 3
Measuring on very steep slopes, from the lowest practical point

Figure 5
Measuring multi-stemmed trees

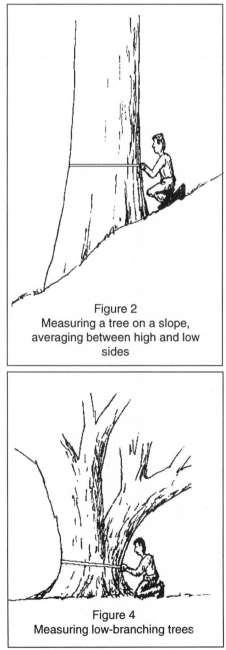

Figure 2
Measuring a tree on a slope, averaging between high and low sides

Figure 4
Measuring low-branching trees

Height

Tree heights are much more difficult to measure. Accurate measurements require the use of fairly sophisticated equipment, such as a relascope or transit. Hand-held devices such as an Abney level or clinometer also give good results when care is taken. All measurements in this book have been taken with one of these methods. These methods require finding a spot where both the top and bottom of the tree are visible. The angle to the top (α_t) and bottom (α_b) is recorded, along with the distance (d) to the tree. The height of the tree on flat ground is then:

$$\text{Height} = d * [tan\,(\alpha_t) + tan\,(\alpha_b)] \qquad\qquad (1)$$

On slopes the distance measured to the tree is not the perpendicular distance, thus a cosine correction must be added (Fig. 6). The height formula then becomes:

$$\text{Height} = [cos\,(\alpha_b) * d] * [tan\,(\alpha_t) + tan\,(\alpha_b)] \qquad\qquad (2)$$

Leaning trees are especially difficult to measure. If the tree is leaning only slightly, the problem can be avoided by measuring the height perpendicular to the lean. Strong leans, however, cannot be

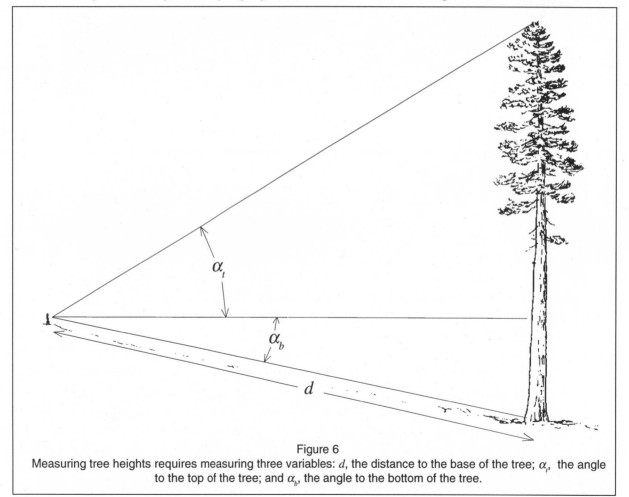

Figure 6
Measuring tree heights requires measuring three variables: d, the distance to the base of the tree; α_t, the angle to the top of the tree; and α_b, the angle to the bottom of the tree.

measured in this way, so the lean must be measured. The formulas presented above assume that the top of the tree is directly above the base of the tree. If this is not the case the tree will be calculated to be either taller or shorter then it really is, depending on the location of the observer. For example, consider the tree depicted in Figure 7. If the observer were to measure the tree from location A and use formula (1), the tree's height would be calculated to be equal to line segment $\overline{Bc}$ – a line much shorter than the real height, $\overline{Bb}$. Alternatively, an observer at D would see the top of the tree as being at a, thus calculating the tree's height as being larger than it really is ($\overline{Ba}$).

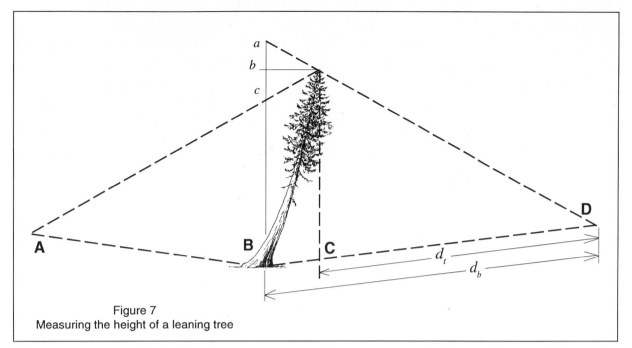

Figure 7
Measuring the height of a leaning tree

The way to account for the lean is to measure two distances, one to the base of the tree (d_b), and the other to a point below the tip of the tree (d_t), at **C**. The formula for calculating the height of a leaning tree is then:

$$\text{Height} = [cos\,(\alpha_b)] * \{[(d_b * tan\,(\alpha_b)] + [(d_t * tan\,(\alpha_t)]\} \qquad (3)$$

To calculate the total length of the tree, L, the distance from the base of the tree to the spot directly under the tip, $\overline{\text{BC}}$, is used with the height using the Pythagorean theorem:

$$L = \sqrt{\text{HT}^2 + \overline{\text{BC}}^2} \qquad (4)$$

Height Estimation

Most people don't have access to sophisticated devices, however, and must make an estimate. Estimated trees are remeasured before inclusion in the register. One simple way to estimate is to use a straight rod, such as a dowel or yardstick, with a 100' tape measure. The method uses the law of similar triangles to calculate the height. Locate yourself about as far away from the tree as you think the tree is tall, hold the rod vertical at arm's length, with the length of the rod above your hand equal to the distance from your hand to your eye, and line up the base of the tree with the top of the hand holding the rod. If the top of the stick appears above the top of the tree, walk forward until they are equal. If the top of the tree is higher than the top of the rod, back up until they are aligned. When the length of the rod at arm's length appears to be the same height as the tree, your distance from the tree is equal to the tree's height. As mentioned above, many trees lean, and it is important to take this into account when measuring. If the tree leans slightly, it is best to measure from a direction perpendicular to the lean. If the tree leans substantially, the lean must be measured. In this case, first locate a point on the ground directly below the tip of the tree. The length of the tree is then calculated as above.

Volume Measurements

A few of the largest trees (e.g., Sitka Spruce, Douglas-fir) have volume figures included with the other measures. Due to the unusual size of the trees and the influence of roots when measuring the circumference at breast height, these trees are best compared by using the amount of wood in the trunk. The volumes are measured using a Criterion 400 Survey Laser.

Crown Spread

This measurement is the average spread of branches, taken as the average between the widest spread and the narrowest spread. This is done by standing under the farthest reaching branch on one side and measuring to the farthest reaching branch on the opposite side. This is then repeated with the narrowest spread (Fig. 8). The two measurements are then averaged. The average crown spread is basically the average diameter of the crown. This is most easily done with an assistant and a 100' tape measure. An alternate method that yields basically the same result is to measure any two perpendicular crown diameters and average them.

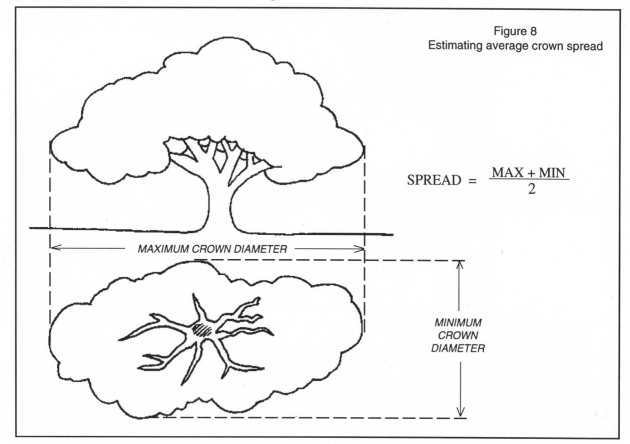

Figure 8
Estimating average crown spread

$$\text{SPREAD} = \frac{\text{MAX} + \text{MIN}}{2}$$

MAXIMUM CROWN DIAMETER

MINIMUM CROWN DIAMETER

AFA Points

In determining bigness, the AFA and most states have adopted a point system giving one point for each inch in circumference at breast height (4 ½ feet), one point for each foot in height, and one point for every 4 feet of average crown spread. For example, the national champion Western Paper Birch in Bellingham is 15' 9" in circumference (189 points), 81' tall (81 points), and has an average crown spread of 57' (14 points), for a total of 284 points. This system has been chosen for the Washington program because of its widespread use, but it does have some shortcomings. It does not record the maximum of any one dimension, such as the tallest height a species has reached. The champion Douglas-fir, for example, is only 205' tall, but there are several in the state over 300', including one 326' tall. Therefore, most entries have more than one tree listed, and the dimensions in bold type are the maximum presently known for that species in Washington. Co-champion status is awarded to trees of nearly the same size (within 5 percent of their total point value). The diamond star (❖) denotes a National Champion tree – the largest known in the United States.

Nominating a Tree

Use the form (or a photocopy) on the last page of this book to nominate a big tree. Fill it out completely, and send it along with a photograph of the tree to the address at the bottom of the form. If you are unsure of the exact species, contact a local expert, or send in a twig and leaf specimen along with a photograph for identification.

Scale is sometimes diffucult to comprehend among the rain forest giants of the Olympic Peninsula. Human scale is essential to judge the size of the **Queets Spruce** – the world's largest known spruce – over 12,000 cubic feet of wood in the trunk alone. Estimated to be between 700 and 800 years old, this tree is still rapidly growing. It is located in the rain forest at the Queets River campground in Olympic National Park.

The Queets fir – the largest known living **Douglas-fir** tree. This tree is a mere shell of its former self. The top is broken out at 205', leaving only the lowest system of branches – it is now basicaly a huge snag with a few branches at the top. The tree was once probably nearly 300 feet tall, based on the top diameter of 6.7 feet. Other than several Coast Redwoods and Giant Sequoias, there is currently no other tree on earth with a diameter over six feet at 200 feet off the ground!

Champion Trees
of Washington State

REGISTER OF WASHINGTON STATE BIG TREES

NATIVE TREES

Circumference	Height	Crown Spread	AFA Points	Date Last Measured	Location and Nominators
MOUNTAIN ALDER					*Alnus tenuifolia*
7'10"	71'	39'	❖175	1993	Umatilla NF, N Fork Asotin River, mouth of Horsethief Canyon *Slim Stillman*
RED ALDER					*Alnus rubra*
16'3"	73'	76'	**287**	1990	Old Nisqually, 639 Old Pacific Hwy SE *RVP (photo below)*
14'10"	86'	**81'**	284	1987	Seattle, Golden Gardens Park *ALJ*
4'2"	**136'**	30'	193	1989	Olympic National Forest, Hamma Hamma River Valley *R Lescher*
SITKA ALDER					*Alnus sinuata*
2'9"	30'	39'	❖73	1992	Maury Island, 24029 59th Ave SW *Mike Lee*
2'3"	**37'**	29'	❖71	1992	Maury Island, 24029 59th Ave SW *Mike Lee*
WHITE ALDER					*Alnus rhombifolia*
9'10"	67'	67'	202	1990	Walla Walla, Ft Walla Walla Natural Area *Shirley Muse*
OREGON ASH					*Fraxinus latifolia*
19'3"	68'	55'	313	1990	Beacon Rock State Park, prairie W of Woodward Ck *RVP (photo page 2)*
15'3"	83'	**82'**	286	1990	Woodland, ½ mile NE of Lewis River bridge *RVP*
5'10"	**111'**	39'	191	1990	Tacoma, Wapato Park *RVP*
QUAKING ASPEN					*Populus tremuloides*
5'5"	**95'**	22'	165	1995	Newport, Cabin Creek Ranch on Trask Pond Rd *Jeri Cross, Ruby Niemeyer*

Near the Nisqually River, 2 miles from where it empties into Puget Sound, stands the state's largest known **Red Alder**. Over five feet in diameter and 73 feet tall, this tree may be found in the pioneer town of Old Nisqually on the Old Pacific Highway SE.

The nation's largest known **Western Paper Birch** lives in Bellingham. This tree has been recognized since 1988. The Western Paper Birch is the western race of the Paper Birch that grows throughout the northern United States and Canada.

Western Paper Birch

NATIVE TREES

PACIFIC BAYBERRY
Myrica californica

| 3'1" | 29' | 25' | 72 | 1994 | Seattle, Seattle Pacific University *ALJ* |

WATER BIRCH
Betula occidentalis

| 4'10" | 45' | 45' | 114 | 1993 | Walla Walla, Whitman College *Shirley Muse, RGB, RVP* |
| 3'10" | 55' | 39' | 111 | 1993 | Walla Walla, Whitman College *Shirley Muse, RGB, RVP* |

1 of 2 trunks

WESTERN PAPER BIRCH
Betula papyrifera var. *commutata*

15'9"	81'	57'	❖284	1989	Bellingham, 4016 Northwest Dr *RVP* (photo page 1)
13'5"	87'	71'	266	1995	Arlington, 13911 Jordan Rd *George Simpson*
8'2"	94'	50'	204	1992	Sedro Woolley, Northern State Multi Service Center *RGB, RVP*

CASCARA
Rhamnus purshiana

| 8'8" | 51' | 43' | ❖166 | 1992 | Gold Bar, 100 yds E of Green Water Meadow Rd on SR 2, E of town *RGB* |
| 2'10" | 70' | 20' | 109 | 1988 | Blaine, 3575 H St *Roger Boettcher* |

Beacon Rock State Park

Located in the heart of the Columbia Gorge, Beacon Rock State Park is named for a giant volcanic plug which rises 500 feet above the Columbia River. Long extinct, this former volcano is an interesting geologic feature which suggests the volcanic origins of our Cascade Mountains. The park is also home to a very rare ecosystem in the Northwest - the oak-ash savanna. This large prairie is located west of Woodward Creek, on the banks of the Columbia River, and is home to a state record tree and two national champion trees. The state record is our native **Oregon Ash**, over six feet in diameter and located in the middle of the prairie. The two national record trees are much smaller than the ash but are the largest known in the country: **Black Hawthorn** and **Western Serviceberry**. The park is worth a visit not only for its namesake feature, but for the prairie remnant which has scattered oaks, ashes, and cottonwoods, along with record trees.

Western Serviceberry

The **Oregon Ash** is a floodplain and swamp tree common throughout the southern Puget Sound area, as well as the Chehalis and Cowlitz river basins. This specimen at Beacon Rock State Park is the largest known in Washington.

The **Alaska Cedar** is one of the longest-lived trees native to the Pacific Northwest. Many fairly small trees have been shown to be over 1000 years old! The oldest verified tree was found as a stump in a clearcut north of Vancouver, BC. It was 1693 years old when cut. The tree pictured, growing above Quinault Lake on the Big Creek Trail in Olympic National Park, is the world's largest known. At twelve feet in diameter, this tree possibly has been alive for over two millennia. Found by Robert Wood and John Aho in 1979, this rain forest giant has the rugged character of its larger cousin, the Western Redcedar.

NATIVE TREES

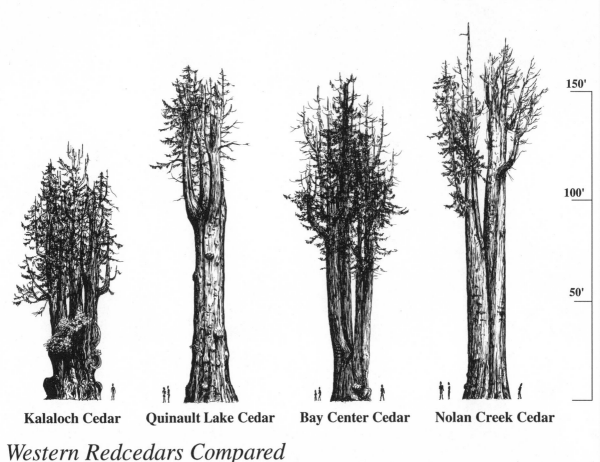

Kalaloch Cedar **Quinault Lake Cedar** **Bay Center Cedar** **Nolan Creek Cedar**

150'

100'

50'

Western Redcedars Compared

The AFA Point system does not work well with large cedars. Very large trees with large diameters accumulate so many points that the AFA Point system inadequately describes the size of a tree. Volume is a much more useful measure but is often difficult to calculate. Measuring the volume of trees with single stems, such as firs and spuces, is fairly straightforward, but trees with big, burly trunks and irregular branching patterns can make volume calculations difficult. Here are silhouettes of the four largest cedars in Washington for comparison.

Nolan Creek Cedar

Giant **Western Redcedars** were once common along the Washington coast. Today few remain. One of the most impressive is on State land south of Forks. Discovered in 1978 when the Department of Natural Resources was cruising for a timber sale along Nolan Creek, the tree was spared but the remaining forest clearcut. This area is now a forest of tiny trees growing up around a behemoth 19 feet in diameter (*see photo, page 106*). While many other trees would die from a dramatic change of environment such as this tree has experienced, Redcedars are adapted to such changes.

Coastal Redcedars grow in areas that experience high winds and every few centuries or so the forest blows over - except for the Redcedars. The species can live for well over one thousand years. Their longevity, combined with their massive base and decay-resistant wood, make ancient Redcedars nearly indestructible. The Nolan Creek champion has probably survived a number of blowdowns in its lifetime. The tree now stands like a ghost over the surrounding plantation, its massive trunk bleached and nearly dead save for a small strip of live bark growing on one side. Ironically, its one live root was cut to install a walkway around the tree designed to protect it! With the fate of this tree in question, Washington has several other contenders in the wings waiting to take the crown of our country's largest Western Redcedar (we can no longer claim to have the world's largest, as a British Columbia tree is larger).

4

The National Co-Champion **Western Redcedar** dwarfs the author in this late summer photo. Its twenty foot diameter trunk is the largest chunk of wood in the state. This tree is near the North Shore Road at Quinault Lake and was the National Champion during the 1940's. Its height was improperly measured however, and the tree was dethroned first by the Kalaloch tree and then the Nolan Creek tree. A remeasurement in 1994 allowed it to regain its record status, as did the volume measurements in 1996.

NATIVE TREES

ALASKA-CEDAR · *Chamaecyparis nootkatensis*

Circumference	Height	Crown Spread	AFA Points	Date Last Measured	Location and Nominators
37'7"	124'	27'	❖582	1994	Olympic NP, Big Creek Tr (4,533 ft³) *Robert Wood and John Aho (photo page 3)*
16'5"	**189'**	33'	394	1988	Gifford Pinchot NF, FS road 21, 12 mi S of US Hwy 12 *RVP*

WESTERN REDCEDAR · *Thuja plicata*

Circumference	Height	Crown Spread	AFA Points	Date Last Measured	Location and Nominators
63'5"	159'	45'	❖931	1993	Olympic NP, N shore of Quinault Lake (14,543 ft³) *F.W. Mathias (photo page 5)*
61'0"	178'	54'	❖924	1977	Forks, Nolan Creek Rd S of Hoh R (13,870 ft³) *Ken Hoover (photo page 106)*
64'2"	123'	51'	**906**	1990	Olympic NP, Hwy 101 N of Kalaloch (12,397 ft³) *F. Dickinson (photo page 22)*
58'3"	154'	55'	867	1993	Bay Center, 4 miles E on River Rd (~11,514 ft³) *Bart Kenworthy, Dick Wilson*
19'9"	**234'**	35'	480	1988	Mt Rainier National Park, Ohanepecosh River and State Hwy 123 *RVP*

BITTER CHERRY · *Prunus emarginata*

Circumference	Height	Crown Spread	AFA Points	Date Last Measured	Location and Nominators
4'10"	**100'**	27'	❖165	1994	Seattle, Seward Park *ALJ*
5'4"	65'	**55'**	143	1992	Tacoma, Parkway W and 29th St W *RVP*

WESTERN CHOKECHERRY · *Prunus virginiana* var. *demissa*

Circumference	Height	Crown Spread	AFA Points	Date Last Measured	Location and Nominators
3'3" largest of 3 stems	41'	31'	**88**	1993	Walla Walla, Ft Walla Walla Natural Area *Shirley Muse, RGB, RVP*
2'5" largest of 4 stems	**44'**	**39'**	83	1993	Walla Walla, Mountain View Cemetery *RGB, RVP*

GOLDEN CHINQUAPIN · *Chrysolepis chrysophylla*

Circumference	Height	Crown Spread	AFA Points	Date Last Measured	Location and Nominators
8'1"	66'	**40'**	**173**	1990	Woodinville, 13538 NE 188th Place *RGB*
5'4"	**72'**	21'	143	1990	Olympic NF, Washington Creek *Ward Willits, Patricia Grover*

BLACK COTTONWOOD · *Populus trichocarpa*

Circumference	Height	Crown Spread	AFA Points	Date Last Measured	Location and Nominators
27'5"	137'	79'	**486**	1995	Sylvania, 1221 Norman Rd *RGB*
25'4"	146'	**88'**	472	1993	Carnation, Carnation-Duvall Rd NE, mile post 10.5 *Andy Weiss*
26'0"	128'	85'	461	1995	Mt Vernon, Jungquist Rd W of Kamb Rd *RVP*
13'7"	**188'**	61'	366	1988	Olympic National Park, Queets campground *RVP*

OREGON CRABAPPLE · *Malus fusca*

Circumference	Height	Crown Spread	AFA Points	Date Last Measured	Location and Nominators
5'5"	**79'**	47'	❖156	1988	Nisqually Wildlife Refuge, along dike trail *RVP*
7'3"	47'	43'	**145**	1992	Fife, Tacoma Country Estates, River Rd & 35th Ave Ct E *RGB, RVP*
5'4"	40'	**54'**	118	1987	Seattle, Washington Park Arboretum, 38N 2E *ALJ*

PACIFIC DOGWOOD · *Cornus nuttallii*

Circumference	Height	Crown Spread	AFA Points	Date Last Measured	Location and Nominators
10'2"	**65'**	**59'**	202	1995	Olympia, 113 17th St *RVP*
10'10"	49'	51'	192	1992	Tacoma, Seaview Dr, W of Orchard *RVP*

WESTERN DOGWOOD · *Cornus occidentalis*

NO DATA

COAST DOUGLAS-FIR · *Pseudotsuga menziesii* var. *menziesii*

Circumference	Height	Crown Spread	AFA Points	Date Last Measured	Location and Nominators
45'5"	205'	37'	**759**	1985	Olympic NP, Kloochman Rock Trail, Queets Fir (~13,616 ft³) *Preston Macy*
40'0"	215'	68'	712	1995	Pe Ell, near Rock Creek (7,415 ft³) *Steve Barnoe-Meyer*
37'4"	298'	64'	**762**	1988	Olympic National Park, S Fork Hoh Trail (7,274 ft³) *RVP, Robert Wood*
39'11"	254'	60'	748	1988	Olympic National Park, Hoh Valley near Happy Four shelter (6,744 ft³) *RVP*
21'1"	**326'**	49'	591	1988	Olympic National Park, Queets Valley near Smith Place *RVP*

ROCKY MOUNTAIN DOUGLAS-FIR · *Pseudotsuga menziesii* var. *glauca*

Circumference	Height	Crown Spread	AFA Points	Date Last Measured	Location and Nominators
19'8"	**173'**	36'	418	1992	Umatilla NF, Turkey Creek Recreational Residence Site *Jim Tardif*

RANDY STOLTMANN PHOTO

Randy Stoltmann is shown standing with the Queets fir, the world's largest **Douglas-fir.** Although much larger trees once existed, this one serves as a partial reminder of times past. Although most of the crown is gone (only one large branch system remains) this is the only Douglas-fir tree that evokes the image of what many of the giants of the past must have looked like. It is also our only example with over 10,000 cubic feet of wood.

The Champion of Firs

The Queets fir was nominated as the largest Douglas-fir in 1945 by Preston Macy, former Olympic National Park Superintendent. It held this title undisputed until September 1962, when the governors of Oregon and Washington assembled a group of judges to measure the mighty Queets fir and Oregon's largest, the Clatsop fir.

Results of 1962 survey	Clatsop Fir, Oregon	Queets Fir, Washington
Diameter at breast height, feet	15.48	14.46
Height to broken top, feet	200.5	202
Diameter at broken top, feet	4.5	6.7
Gross volume outside bark, cubic feet	10,095	14,063

The Oregon tree was crowned champion based on points - 38 points higher than the Queets fir (largely due to a larger basal diameter) although the Queets fir was much larger in terms of volume. The point was moot, however, because two months later the famous Columbus Day storm blew the Oregon tree down.

And so it went for another decade until 1972, when Finnegan's Fir was nominated. Another Oregon tree, much thinner than the Clatsop or Queets firs, Finnegan's Fir had enough points to hold the new record, due to its intact top, 302 feet high. Located in the Coos Bay area, an area famous for firs, it had regained the title for Oregon. This record too, was short-lived. Slightly more than six months after it had been nominated, it fell down. Again the Queets tree regained the title.

In 1988, I was informed by Bob Wood, author of several books on the Olympic Mountains, of a giant he knew about on the South Fork Hoh River. Jerry Franklin, forestry professor at the University of Washington, also knew of the tree and considered it the most perfect Douglas-fir tree he'd ever seen. While much more youthful in appearance than many other large firs, it has all the makings of a champion. Upon measurement it had enough points to match the Queets tree even though it has much less wood - again due to its intact top.

These two Washington trees shared the title for a couple of years until another Oregon tree in the Coos Bay area was discovered. Called the Brummet Fir, it is 326 feet tall. The Brummet Fir is currently recognized as the National Record, although at 8,250 cubic feet, is much smaller than the Queets tree.

NATIVE TREES

Circumference	Height	Crown Spread	AFA Points	Date Last Measured	Location and Nominators
BLUE ELDER					*Sambucus caerulea*
8'7" fused base	47'	31'	**157**	1993	Walla Walla, Ft Walla Walla Natural Area *Shirley Muse, RGB, RVP*
5'2"	**48'**	32'	118	1993	Brush Prairie, NE 82nd Ave, N of E Fk Lewis River *RGB, RVP*
PACIFIC RED ELDER					*Sambucus callicarpa*
4'0"	**36'**	29'	❖**91**	1993	Coupeville, mile marker 25 on State Hwy 20 *RGB*
ALPINE FIR					*Abies lasiocarpa*
21'0"	125'	26'	❖**383**	1992	Olympic National Park, Cream Lake *Stephen Arno & Jeff Hart (photo page vi)*
11'0"	**172'**	23'	310	1988	Alpine Lakes Wilderness, Icicle Creek trail *RVP*
GRAND FIR					*Abies grandis*
19'1"	251'	43'	❖**491**	1988	Olympic National Park, Duckabush River trail (~2,424 ft³) *RVP*
20'4"	217'	47'	**466**	1994	Olympic National Park, Barnes Crk at US 101 (2,264 ft³) *RVP (photo page 24)*
16'3"	**267'**	28'	469	1993	Glacier Peak Wilderness, Suiattle River at Miner's Creek (~1,870 ft³) *RVP*
NOBLE FIR					*Abies procera*
28'4" broken - was 278'	238'	41'	❖**588**	1988	Gifford Pinchot NF, Yellowjacket Ck (~6,153 ft³) *H Coates & B Smith (photo page 21)*
25'0"	272'	49'	❖**584**	1989	Mt St Helens Nat Vol Mnmt, Goat Marsh RNA (4,802 ft³) *RVP (photo page 10)*
19'9"	**295'**	43'	543	1989	Mt St Helens National Volcanic Monument, Goat Marsh RNA (3,096 ft³) *RVP*
PACIFIC SILVER FIR					*Abies amabilis*
24'5"	217'	31'	❖**518**	1989	Forks, Goodman Creek (3,433 ft³) *Lloyd H Larson (photo page 10)*
14'5"	**236'**	35'	418	1989	Olympic National Forest, E Fork Humptulips River *R Lescher*
NETLEAF HACKBERRY					*Celtis reticulata*
3'5"	**25'**	35'	75	1993	Spokane, Finch Arboretum *RGB, ALJ*
BLACK HAWTHORN					*Crataegus douglasii*
9'3" largest of 3 trunks	41'	57'	❖**166**	1993	Beacon Rock State Park, prairie W of Woodward Ck *RVP*
8'4"	49'	50'	**161**	1993	Vancouver, Vancouver Lake Park *RVP*
7'6"	**51'**	42'	152	1988	Quinault, E end of Lake Quinault *Randy Stoltmann*

Alpine Fir Adventure

In the middle of the Olympic Mountains, about as far from any road or trail as one can get, stands the largest known **Alpine Fir**. Located at Cream Lake near the backbone of the Bailey Range, the tree stands in silent testimony to the passage of time. This tree has witnessed fewer human visitors than any other champion tree in our state. The tree was discovered in 1963 as Stephen Arno, then park ranger and later author of *Northwest Trees*, was traversing the range with Jeff Hart. The tree is hollow and once supported a small door to the cavity, giving the tree a bizarre elfin appearance.

During the summer of 1993, a fellow tree enthusiast and I hiked to see the tree and get updated measurements (and to check if it was still alive, since it hadn't been remeasured since 1964!). The lack of trails for most of the way, the ruggedness of the Olympic Mountains, and a blinding fog all combined to thwart our attempts to make this a three-day hike. Late in the afternoon of the fourth day, we dragged ourselves back to the trailhead, weary but successful. Growth rates are low in alpine environments, and we expected little growth, yet could detect none. In fact, we measured it as shrinking one inch in circumference. The door is long gone but the tree is still healthy. The large hollow, however, indicates the tree is supported by a shell of wood only one to two feet thick, and this seemingly immortal tree could come down in a big wind. (See photo on page vi.)

Washington Firs

Washington has four native species of true fir (*Abies*) and has the National Record of each of the four. **Alpine Fir** is found from Alaska to New Mexico and from sea level to over 12,000 feet, perhaps the widest elevational range of any tree. The **Pacific Silver Fir** has a range that defines the Coastal Pacific Northwest Biome. It grows from Alaska to northern California only in moist, cool environments – reaching its maximum development on the wet western side of the Olympic Peninsula. **Grand Fir** is common in the mountains of eastern Washington, Idaho, northern Montana, and south-eastern British Columbia, yet achieves greatest size at low elevations along rivers in western Washington and adjacent British Columbia. **Noble Fir** occurs only in Cascade subalpine forests south of Stevens Pass and a small population in the Willapa Hills. The Noble Fir on Yellowjacket Creek is the largest known specimen of any type of *Abies* in the world.

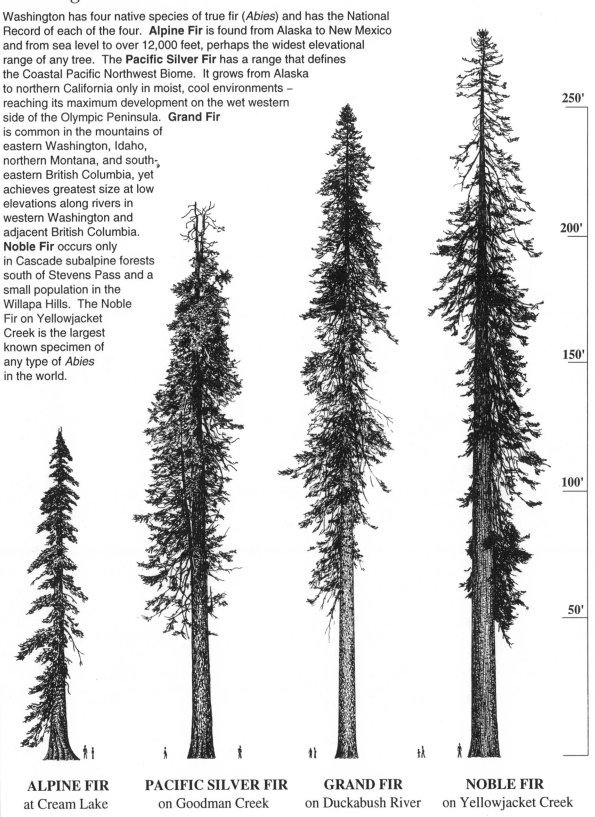

ALPINE FIR	**PACIFIC SILVER FIR**	**GRAND FIR**	**NOBLE FIR**
at Cream Lake	on Goodman Creek	on Duckabush River	on Yellowjacket Creek

250'

200'

150'

100'

50'

NATIVE TREES

COLUMBIA HAWTHORN   *Crataegus columbiana*

2'3"	**18'**	19'	**50**	1993	Spokane, Finch Arboretum *RGB, ALJ*

CALIFORNIA HAZEL   *Corylus cornuta* var. *californica*

2'6"	**41'**	51'	**84**	1990	Seattle, Golden Gardens Park *RVP, ALJ*

one of many stems

2'1"	31'	**56'**	70	1993	Maury Island *Mike Lee, RGB*

MOUNTAIN HEMLOCK   *Tsuga mertensiana*

19'6"	152'	41'	❖**396**	1993	Olympic NP, Enchanted Valley, O'Neil Pass trail *Robert Wood (photo page 11)*
16'1"	174'	49'	**379**	1988	Olympic National Park, Wynoochee trail *RVP*
13'1"	**194'**	29'	358	1988	Olympic National Park, near Sundown Pass *RVP*

WESTERN HEMLOCK   *Tsuga heterophylla*

26'4"	202'	47'	❖**530**	1988	Olympic National Park, Wynoochee trail (3,588 ft³) *Robert Wood, RVP*
27'3"	164'	37'	**508**	1993	Glacier Peak Wilderness, Milk Creek trail (~3,429 ft³) *Randy Stoltmann*
28'5"	174'	65'	❖**531**	1993	Olympic NP, Enchanted Valley (3,072 ft³) *Preston Macy (photo page 11)*
22'6"	**241'**	68'	❖**528**	1988	Olympic National Park, Hoh River trail (2,029 ft³) *RVP*

ROCKY MOUNTAIN JUNIPER   *Juniperus scopulorum*

12'1"	62'	30'	**214**	1994	Skagit Island *Chris Chappell*
9'8"	**64'**	28'	187	1988	Anacortes, Washington Park near Fidalgo Head *RVP*

Although the species is normally found at 3,000 – 6,000 feet in the Cascades, this giant **Pacific Silver Fir** grows on the coastal plain of the Olympic Peninsula at 250 feet above sea level. Found on State land in the Goodman Creek drainage south of Forks, this 217' giant looms over the stumps of its former neighbors.

My brother Bruce is shown admiring not only the National Co-Champion **Noble Fir**, but the densest forest stand in the state, found on the southern slope of Mount Saint Helens. Containing up to 60,000 cubic feet of wood per acre, this patch of woods is exceeded in density only by some Coast Redwood and Giant Sequoia stands in California.

Circumference	Height	Crown Spread	AFA Points	Date Last Measured	Location and Nominators

WESTERN JUNIPER   _Juniperus occidentalis_

| 7'6" | 37' | 41' | **137** | 1988 | Juniper Dunes Wilderness, N of Pasco _RVP_ |
| 5'10" | **52'** | 33' | **130** | 1993 | Zillah, Zillah Cemetery _RVP, RGB_ |

ALPINE LARCH   _Larix lyallii_

| 19'8" | **94'** | 56' | ❖**344** | 1972 | Wenatchee National Forest, Big Mtn, Chelan Range _Stephen Arno, Jeff Hart_ |

The Enchanted Valley Hemlocks

Deep in the heart of the Olympic Mountains lies the Enchanted Valley, Washington's version of Yosemite Valley, with its flat, grassy meadows scattered with trees below spectacular mountain walls rising 2,500 feet from the valley floor. Also called the Valley of 1,000 Waterfalls, Enchanted Valley has fascinated visitors for over a century. At the head of the valley stands one of the largest known **Western Hemlocks**. Nine feet in diameter, this tree has been on the books since Preston Macy nominated it in 1945. Briefly dethroned in the late 1980's by the other National Co-Champions listed in this book, it was remeasured in August of 1993 and had grown enough to reclaim its title.

Less than a mile away, at 4,000 feet, stands Washington's largest known **Mountain Hemlock**. Over six feet in diameter and 152 feet tall, it is also a National Co-Champion and among the largest of its species in the world. Mountain hemlocks grow at snowy high elevations in the Pacific Northwest; their graceful windswept branches adorn mountaintops throughout the Cascade and Olympic Mountains. Both trees may be close to a millennium in age and represent the ultimate development of their genus.

The state tree of Washington is **Western Hemlock**. Ironically, there are currently three trees considered co-champions by the American Forestry Association as the largest known in the United States. All three grow in Olympic National Park, and this one, in the Enchanted Valley, has the largest known diameter.

Mountain Hemlocks adorn the high ridges of the Olympic, Cascade, and Sierra Nevada Mountains. They also occur in the Coast Ranges of British Columbia and Alaska. The largest in Washington is an ancient beast found high in the rainforest of the Quinault River above the Enchanted Valley.

Another view of one of our largest **Pacific Madrona.** Although disease is threatening a large proportion of our Madrona population and many of our city Madrona trees are looking weak, this specimen remains glorious. Madronas are often located in places where humans like to build houses. While many homeowners like the trees and try to save them, Madronas have delicate roots that are often damaged by housing construction and die anyway. This tree however, seems to have survived constuction of a building on part of its roots. Located on 8th Street in Port Angeles, this tree is worth driving a few blocks from HWY 101 to see.

RANDY STOLTMANN PHOTO

The **Bigleaf Maple** is the largest species of maple in the world. Many giant specimens are found throughout the rural areas of western Washington. The specimen pictured above may be found in the Skagit Valley near Hamilton and is one of several in the state about ten feet in diameter. We're sure this dimension can be beaten. If you know of a larger one in some farmland or wild area near you, be a hero: let us know.

Circumference	Height	Crown Spread	AFA Points	Date Last Measured	Location and Nominators

WESTERN LARCH
Larix occidentalis

19'2"	189'	35'	❖**428**	1993	Wenatchee NF, Pleasant Valley Campground *KVP, RVP (photo page 23)*
14'3"	**192'**	38'	373	1989	Umatilla NF, Green Creek, lower slope of Little Turkeytail Ridge *Jim Tardif*

PACIFIC MADRONA
Arbutus menziesii

23'4"	78'	55'	**372**	1993	Tacoma, Pt Defiance Park *RVP*
21'0"	85'	**95'**	361	1993	Port Angeles, 231 W 8th St *RVP (photo pages iii, 12)*
20'2"	97'	84'	360	1987	Seattle, Seward Park *ALJ*
15'9"	**111'**	85'	321	1987	Seattle, Seward Park *ALJ*

BIGLEAF MAPLE
Acer macrophyllum

33'2"	91'	86'	**510**	1992	Hamilton, Hwy 20 W of Healy Rd *Randy Stoltmann (photo page 12)*
32'11"	91'	**95'**	**510**	1993	Sylvana, 614 Pioneer Hwy *RGB*
14'1"	**158'**	61'	342	1989	Mt Baker NF, 2.5 mi in on FS Rd 25 *RVP*

Many giant **Oregon White Oaks** grow on the Columbia River floodplain near Woodland. The largest is this one, nearly seven feet in diameter, seen here with nominator Byron Ferguson (*right*) and Ron Brightman.

This specimen of **Western White Pine** on the Lewis River in southern Washington is the tallest known pine of any kind in the state. At 214 feet it is impressive, yet much shorter than what this species is capable of reaching.

13

Big Pines

Washington's largest **Ponderosa Pines** are located near Mount Adams in the southern Cascade Range. One is 3 miles north of Trout Lake in the Gifford Pinchot National Forest. The other is on the northeast side of Mount Adams within the Yakama Indian Reservation. The National Forest tree is in a fenced enclosure with a dedicationto Elmer W. Lofgren, a former Forest Service ranger who helped protect the tree. Fifty years ago the tree was blazed for harvest, since it grows in an area with many large pines. Some fellow (some say Lofgren) scraped off the blaze along with some bark to spare the tree. Because of the thick bark of mature Ponderosa Pines, the tree was uninjured. The narrower section of the tree (at breast height) is still visible. The Lofgren tree is seven feet in diameter and scales 3,248 cubic feet of wood. The Yakama tree, while quite a bit shorter, has a larger diameter and slightly more wood: 3,434 cubic feet. These two trees have been known about for years and each group (the Forest Service and Yakama Indian Nation) has claimed theirs as the largest. It wasn't until I took the survey laser out to both trees and measured their volumes that we learned that the Yakama tree has slightly more wood volume. These aren't the world's largest, however: a few trees in California and Oregon are larger.

200'

150'

100'

50'

YAKAMA PINE   **TROUT LAKE PINE**

Circumference	Height	Crown Spread	AFA Points	Date Last Measured	Location and Nominators

DOUGLAS MAPLE — *Acer glabrum* var. *douglasii*

Circumference	Height	Crown Spread	AFA Points	Date Last Measured	Location and Nominators
8'11" below forking	67'	**55'**	❖**188**	1995	Guemes Island, 0.9 miles S of Eden's Rd on West Shore Dr *RGB, RVP*
6'1"	66'	43'	150	1995	Guemes Island, 0.9 miles S of Eden's Rd on West Shore Dr *RGB, RVP*
4'1"	**77'**	29'	133	1995	Guemes Island, 0.9 miles S of Eden's Rd on West Shore Dr *RGB, RVP*

VINE MAPLE — *Acer circinatum*

Circumference	Height	Crown Spread	AFA Points	Date Last Measured	Location and Nominators
2'11"	**62'**	31'	**105**	1988	Olympic National Park, S Fork Hoh trail *RVP*
4'6"	32'	41'	96	1992	Vashon Island, across from 10801 Bank Rd *Mike Lee, ALJ*

CASCADE MOUNTAIN ASH — *Sorbus scopulina*

Circumference	Height	Crown Spread	AFA Points	Date Last Measured	Location and Nominators
2'0"	**13'**	19'	❖**42**	1993	Spokane, Finch Arboretum *RGB, ALJ*

SITKA MOUNTAIN ASH — *Sorbus sitchensis*

NO DATA

OREGON WHITE OAK — *Quercus garryana*

Circumference	Height	Crown Spread	AFA Points	Date Last Measured	Location and Nominators
20'11"	93'	106'	**370**	1993	Woodland, 1123 S Pekin Rd *Byron Ferguson (photo page 13)*
19'10"	89'	**111'**	**355**	1995	Toledo, across from 1066 Tucker Rd *Ronald Ogren*
19'1"	**98'**	76'	346	1990	Woodland, 650 S Pekin Rd *RVP*

LODGEPOLE PINE — *Pinus contorta* var. *latifolia*

Circumference	Height	Crown Spread	AFA Points	Date Last Measured	Location and Nominators
13'1"	43'	44'	**211**	1980	Olympic National Park, Deer Ridge trail *Robert Wood*
5'8"	**105'**	24'	179	1990	Mead, 200 feet NE of Holcomb & Travis Roads *Barry Sattin*

PONDEROSA PINE — *Pinus ponderosa*

Circumference	Height	Crown Spread	AFA Points	Date Last Measured	Location and Nominators
22'7"	179'	60'	**465**	1995	Yakama Indian Reservation, West Fork Klickitat River (3,434 ft³) *RVP*
22'0"	**213'**	55'	491	1995	Gifford Pinchot NF, 3 mi N of Trout Lake (3,248 ft³) *Maynard Drawson*

SHORE PINE — *Pinus contorta* var. *contorta*

Circumference	Height	Crown Spread	AFA Points	Date Last Measured	Location and Nominators
11'6"	101'	37'	❖**248**	1992	Bryant, 35th Ave NE, 1 mile N of Stanwood-Bryant Rd *RGB*

Spruce Wars

In December of 1987 five people in two vehicles traveled over 500 miles and measured two **Sitka Spruce** trees along the rugged Pacific Northwest Coast to determine the world's largest spruce. One tree, near Seaside, Oregon, had been the champion for a decade and the other, at Quinault Lake on the Olympic Peninsula in Washington was the contender. The controversy started because of the butress-like root swell that large Sitka Spruce trees develop to stabilize themselves. The AFA, which keeps records on big trees, requires the measurement to be at 4.5 feet above average ground level, yet the influence from roots on trees in the Pacific Northwest may extend well past that point. At the end of the weekend the trees were declared National Co-Champions, which I think is fair given the drawbacks of the point system and the impressive dimensions of both trees.

The battle over which spruce was larger illustrates the shortcomings of using an overall point value for a tree's 'bigness'. Yet I list the Quinault tree second in our state list – the Queets campground tree being first. Why? The *volume* of a tree more accurately represents a tree's size than a point system and, when possible, I will use this. The Queets tree has more volume than either spruce mentioned above, and more than any other spruce! The problem with relying on this method lies in the extreme difficulty in measuring volume versus the relatively easy method for determining points. The point system is a useful one - especially for broadleaf trees that don't have a single straight stem as do many conifers – accurately calculating volume for a wide-spreading oak for example, would be nearly impossible.

The National Co-Champion **Sitka Spruce** dwarfs the author in this springtime photo. This giant tree grows at the east end of Quinault Lake on the western Olympic Peninsula.

Circumference	Height	Crown Spread	AFA Points	Date Last Measured	Location and Nominators

WESTERN WHITE PINE _Pinus monticola_

16'3"	192'	37'	**396**	1988	Alpine Lakes Wilderness, Icicle Creek trail _RVP_
14'6"	**214'**	28'	**395**	1989	Gifford Pinchot NF, mile marker 33 on FS Rd 90 _RVP_ _(photo page 13)_

WHITEBARK PINE _Pinus albicaulis_

16'8"	67'	51'	**280**	1993	Alpine Lakes Wilderness, 4th Ck _Dave Braun, Becky Hudson (photo page 105)_
12'3"	**90'**	29'	244	1993	Alpine Lakes Wilderness, 4th Ck _Dave Braun, RVP_
12'7"	60'	**60'**	226	1987	Mt Adams Wilderness, Avalanche Valley _Jim Riley_

PACIFIC RHODODENDRON _Rhododendron macrophyllum_

NO DATA

WESTERN SERVICEBERRY _Amelanchier alnifolia_

3'3"	**42'**	**43'**	❖**92**	1993	Beacon Rock State Park, prairie W of Woodward Ck _RVP_
largest stem					

ENGELMANN SPRUCE _Picea engelmannii_

19'1"	**213'**	25'	**450**	1987	Alpine Lakes Wilderness, Icicle Creek trail (~1,918 ft³) _RVP_
19'9"	186'	27'	**432**	1987	Alpine Lakes Wilderness, Icicle Creek trail (~1,890 ft³) _RVP_
22'1"	179'	27'	**451**	1995	Olympic National Park, Cameron Creek trail (1,497 ft³) _ALJ, RVP_

The **Engelmann Spruce**, while common throughout the Cascade and Rocky Mountains, also has three small populations in the Olympics. The tree pictured is the largest one known in the Olympics. Located near Cameron Creek, the tree was found last year on an expedition to find a new National Champion tree. Although not big enough, it is nonetheless a new State Champion.

The _tallest_ known **Engelmann Spruce** in the world, however, is in Washington. This 213' tall giant is growing in a beaver swamp along Icicle Creek in the Alpine Lakes Wilderness.

QUEETS SPRUCE   **QUINAULT LAKE SPRUCE**   **PRESTON MACY TREE**

The largest **Sitka Spruce** trees in Washington. These are three of the four known Sitka Spruce trees having over 10,000 cubic feet of wood each. The fourth (third largest in the world) is in Oregon (see story on page 15). The Queets Spruce is not only the largest of its kind, but is one of the fastest-growing trees on earth, adding as much as 700 cubic feet per year based on readings from increment corings. The Quinault Lake Spruce is the current National Co-Champion (with the Oregon tree). Its large root flare gives it more AFA points even though it is slightly smaller. The Preston Macy tree is named after the first superintendent of Olympic National Park, and was the National Champion during the 1940's and 50's.

# of Items:1 Shipping Method: STANDARD

Locator Code	SKU	Title
	A44-4958	Champion Trees of ..

Your Order From: **Amazon**

Order Number: **107-7391366-4200223**

DATE ORDER RECEIVED

August 25, 2014

Order Date: 08/24/2014

FROM: Goodwill Industries of the Columbia
220 W. Columbia Street
Pasco, WA 99301

TO: lk Icard
10836 NE HOOT OWL WAY
KINGSTON, WA 98346
USA

For Customer Service, please contact us at:
ebooks@goodwillotc.org

LOCATION: TYPE: Paperback TITLE: Champion Trees of Washington St ISBN/UPC No.: 9780295975634 INTERNAL SKU No.: A44-4958 Description: A copy that has been used, but remains in clean condition. All pages are intact, and the cover is, this may or may not include dustcover. The spine may show signs of wear. Pages can include limited notes and highlighting, and the copy can include "From th	8.06
Amazon's S&H	3.99
TOTAL Collected by Amazon	12.05

Our Guarantee: We appreciate your business and are committed to providing the best possible service. If you are unhappy with your purchase, please contact us immediately at ebooks@goodwillotc.org and if we are unable to correct the problem to your satisfaction, we will refund your purchase.

Thank you for your order!

Circumference	Height	Crown Spread	AFA Points	Date Last Measured	Location and Nominators

SITKA SPRUCE   *Picea sitchensis*

45'3"	245'	83'	809	1995	Olympic NP, Queets campground, Queets Spruce (12217 ft³) *RVP (photo page xi)*
58'11"	191'	96'	❖**922**	1995	Olympic Nat Forest, SE shore of Quinault Lake (11969 ft³) *RVP (photo page 16)*
41'2"	257'	67'	768	1995	Olympic National Park, Hoh Road, Preston Macy Tree (10852 ft³) *Robert Wood*
21'11"	**305'**	62'	583	1988	Olympic National Park, Queets Valley near Smith Place *RVP*

SMOOTH SUMAC   *Rhus glabra*

| 3'2" | 26' | **26'** | ❖**71** | 1993 | Walla Walla, 20 Merriam St *Shirley Muse, RGB, RVP* |
| 2'2" | **38'** | 19' | ❖**69** | 1993 | Walla Walla, Whitman Mission National Historic Park *RGB, RVP* |

ARROYO WILLOW   *Salix lasiolepis*
NO DATA

BEBB WILLOW   *Salix bebbiana*
NO DATA

HOOKER WILLOW   *Salix hookeriana*

| 3'2" | 28' | 43' | 77 | 1991 | Tacoma, N 21st & Mason Ave *KVP, RVP* |

Ron Brightman stands in awe of the National Champion **Scouler Willow**. Ron found this tree in front of the Ordway Elementary School in Winslow. Only a few minutes from the ferry to Seattle, this tree is an ideal schoolyard tree. Science classes visit to learn about trees. The Scouler Willow is the largest type of pussy willow, and the only one of our native pussy willows to be commonly found away from wet places.

NATIVE TREES

Circumference	Height	Crown Spread	AFA Points	Date Last Measured	Location and Nominators

PACIFIC WILLOW
Salix lasiandra

10'7"	40'	**67'**	**184**	1993	Auburn, across from 509 3rd St SW *RGB*
8'1"	66'	57'	**177**	1993	Gig Harbor, Rosedale St, E of Hwy 16 overpass *RVP*
6'6"	**79'**	40'	167	1993	Seattle, Washington Park Arboretum, K8 *ALJ*

PEACHLEAF WILLOW
Salix amygdaloides

8'0"	**56'**	**95'**	**176**	1993	Walla Walla, Whitman Mission National Historic Park *RGB, RVP*

largest of 6 trunks

PIPER WILLOW
Salix piperi

4'5"	**30'**	**48'**	**95**	1994	Seattle, Washington Park Arboretum, Foster Island *ALJ*

SCOULER WILLOW
Salix scouleriana

14'7"	50'	43'	❖**236**	1995	Winslow, Ordway Elementary School *RGB*
12'0"	64'	47'	**220**	1993	Maury Island, private residence *RGB, RVP*
below forking					
2'8"	**82'**	21'	119	1988	Anacortes, Washington Park *RVP*
8'0"	69'	**53'**	178	1988	Anacortes, Washington Park *RVP*

SITKA WILLOW
Salix sitchensis

1'5"	**36'**	**19'**	**58**	1995	Seattle, Green Lake Park *ALJ*

PACIFIC YEW
Taxus brevifolia

15'0"	54'	30'	❖**241**	1988	Gifford Pinchot NF, Silver Creek *B Malcolm, L Barnhouse, A Storkman, R Levitt*
7'0"	**64'**	43'	159	1988	Seattle, Seward Park *ALJ*
10'2"	38'	**55'**	174	1988	Seattle, Lincoln Park *ALJ*

Jim Riley, Gifford Pinchot National Forest ranger, sits in front of the largest **Pacific Yew**. Made famous overnight after the discovery of the cancer medicine Taxol found in its bark, this little-understood species became nationally recognized. Normally a small understory tree in old-growth forests, this five-foot diameter specimen demonstrates the great size a Pacific Yew can attain given enough time.

QUICK REFERENCE
WASHINGTON'S LARGEST TREES

What criteria do we use to compare two trees to see which is largest? One might have a large trunk but not be very tall. Another may be tall but with a slender trunk. Which one is bigger? The American Forestry Association established a point system designed to compare similar trees. It is a useful system for comparing most of the different kind of trees that we grow (see discussion in the Introduction). Diameter is what most people use to compare trees. The diameter is the part of trees that people can see and easily measure. However, with the giant trees that grow on the Pacific Coast, a diameter measured at breast-height (dbh) will often represent the giant root structure at the bottom of the tree and not the main stem. The actual amount of wood contained in the tree, however, is a true gauge of the growth of trees, although very difficult to measure. Here are the ten largest tree species in Washington based on these three methods.

	Wood Volume (ft³)		AFA Points		Diameter (ft)	
1	Western Redcedar	14,543	Western Redcedar	931	Western Redcedar	20.4
2	Douglas-fir	13,616	Sitka Spruce	922	Sitka Spruce	18.8
3	Sitka Spruce	12,217	Douglas-fir	762	Douglas-fir	14.5
4	Noble Fir	6,153	Noble Fir	588	Alaska Cedar	12.0
5	Alaska Cedar	4,533	Alaska Cedar	582	Bigleaf Maple	10.5
6	Western Hemlock	3,588	Western Hemlock	531	Giant Sequoia *	10.3
7	Ponderosa Pine	3,434	Giant Sequoia *	531	Lombardy Poplar *	9.7
8	Pacific Silver Fir	3,433	Lombardy Poplar *	524	Western Hemlock	9.0
9	Grand Fir	2,424	Pacific Silver Fir	518	Noble Fir	9.0
10	Engelmann Spruce	1,918	Bigleaf Maple	510	Black Cottonwood	8.1

* Introduced species

Noble fir

Jim Riley again, this time with the largest **Noble Fir**. This tree is not only the largest known Noble Fir, but is the largest true fir (*Abies*) in the world. The tree was in a Noble Fir Botanical Area during the 1950's, but the grove, excepting this tree, has since been logged. Recently, the top 40-foot section has blown off.

Iris sits at the base of the Kalaloch Cedar, a massive **Western Redcedar** in Olympic National Park. The tree has the largest known trunk diameter of any of its species. A National Champion during the 1960's, it was dethroned in 1977 by the Nolan Creek tree, which is 55 feet taller. Although there is still some question as to whether it is one tree or three fused trunks, the tree is impressive nonetheless. It may be found about 5 miles north of Kalaloch Beach near Hwy 101 in Olympic National Park's coastal strip.

QUICK REFERENCE
WASHINGTON'S
TALLEST TREES

Douglas-fir	326'
Sitka Spruce	305'
Noble Fir	295'
Grand Fir	267'
Western Hemlock	241'
Pacific Silver Fir	236'
Western Redcedar	234'
Western White Pine	214'
Ponderosa Pine	213'
Engelmann Spruce	213'
Mountain Hemlock	194'
Western Larch	192'
Alaska-Cedar	189'
Black Cottonwood	188'

The National Champion **Western Larch**. Discovered in 1993 on a spring outing, this 189-foot giant towers over the American River in the Wenatchee National Forest. The Pleasant Valley campground features many lofty specimens of Western Larch, including this noble tree.

Western Washington has many large **Black Cottonwoods** – a native poplar common along our river floodpains. Unfortunately they are short-lived, both through natural and human causes. Our two largest were cut down in 1995. This tall tree at a Renton office was over 7 feet in diameter and 160 feet tall when removed in 1995. We are looking for replacement champions of this species.

Ivan stands in front of a giant **Grand Fir** in Olympic National Park. This tree has the largest diameter of any of its species in Washington. Recognized as special by the highway department, the tree was spared when U.S. Hwy 101 was rerouted over Barnes Creek at Lake Crescent. It now stands a few feet from the highway near where it passes over the creek. The nearby forest contains many large Grand Firs, including a former National Champion shown to F.D.R. when he was considering the formation of Olympic National Park.

INTRODUCED TREES

	Circumference	Height	Crown Spread	AFA Points	Date Last Measured	Location and Nominators

ALDER

European *Alnus glutinosa*

| | 5'5" | 65' | 41' | 140 | 1993 | Seattle, Washington Park Arboretum, 13N 1W *RVP* |

Italian *Alnus cordata*

| | 5'7" | 81' | 37' | 157 | 1993 | Seattle, Washington Park Arboretum, 13N 0 *RVP* |
| | 5'6" | 59' | 41' | 135 | 1993 | Seattle, Washington Park Arboretum, 50N 11E *RVP* |

Royal *Alnus glutinosa* 'Imperialis'

| | 4'2" | ~25' | 27' | ~82 | 1993 | Hoquiam,   *RGB* |

ALMOND

Prunus dulcis

| | 4'7" | 32' | 37' | 96 | 1993 | Tacoma, 2336 Tacoma Ave S *KVP, RVP* (photo below) |

Double-flowered Hybrid *Prunus* x *persicoides* 'Roseoplena'

| | 5'10" | 30' | 29' | 107 | 1992 | Seattle, 3415 Cascadia *RGB* |

Hall's Hardy (?) *Prunus* x *persicoides* 'Hall's Hardy'

| | 3'10" | 29' | 33' | 83 | 1993 | Marysville, 1209 Fifth St *RGB* |

APPLE

Common *Malus* x *domestica*

	10'8"	38'	47'	178	1993	Washougal, 27th St, N of Hwy 14 *RGB, RVP*
	8'7"	46'	47'	161	1990	Tacoma, 1703 Alaska St *KVP, RVP*
	8'7"	39' topped	56'	156	1992	Orting, Washington Old Soldiers Home *RGB*

Gravenstein *Malus* x *domestica* 'Gravenstein'

| | 8'3" | 45' | 58' | 158 | 1993 | Snohomish, 9929 Airport Way *RVP* |

The **Almond** is a small, attractive tree uncommonly seen. In flower, it is one of our more beautiful trees, coming into bloom in March. The single, white flowers with a reddish-pink center are followed by simple, lance-shaped leaves. Almonds are generally smaller growing than cousins such as cherries and plums, but they are often healthier looking, as this specimen in Tacoma illustrates.

INTRODUCED TREES

Circumference	Height	Crown Spread	AFA Points	Date Last Measured	Location and Nominators

APRICOT
Prunus armeniaca

8'10"	~50'	~50'	~168	1986	Wishram *Maynard Drawson*

ARALIA

Castor
Kalopanax septemlobus

5'9"	50'	51'	132	1993	Seattle, Washington Park Arboretum, 13N 1W *RVP*

ARBORVITAE   (see also **CEDAR**, page 34; and **CYPRESS**, page 45)

Columnar
Thuja occidentalis 'Fastigiata'

11'1" fused base	38'	23'	**177**	1993	Toppenish, 715 Lincoln Ave *RGB, RVP*
10'6" fused base	38'	27'	**171**	1993	Toppenish, 715 Lincoln Ave *RGB, RVP*
5'10"	**43'**	12'	116	1988	Sedro Woolley, Wicker Rd Cemetery

Douglas
Thuja occidentalis 'Douglasii Pyramidalis'

4'9"	**55'**	25'	118	1993	Walla Walla, 1121 Alvarado St *RGB, RVP*

George Peabody
Thuja occidentalis 'Lutea'

4'0"	**40'**	29'	96	1990	Puyallup, Puyallup High School *RVP*
4'0"	38'	23'	92	1993	Parkland, Pacific Lutheran University *RVP*

Gold Column
Thuja orientalis 'Elegantissima'

4'4"	**37'**	35'	98	1993	Seattle, Leschi Park *ALJ, RVP*

Golden Spike
Thuja occidentalis 'Aureospicata'

6'7"	**55'**	20'	139	1992	Burlington, SW corner of Maiben Park *RGB*
6'10"	50' broken top	24'	138	1992	Burlington, east edge of Maiben Park *RGB*

Golden Western
Thuja plicata 'Aurea'

4'0"	**57'**	23'	111	1993	Seattle, University of Washington, Medicinal Herb Garden *RGB*

Hiba
Thujopsis dolabrata

3'11"	**47'**	22'	99	1989	Tacoma, Old Tacoma Cemetery *RGB, ALJ*

Hybrid
(?) *Thuja occidentalis* x *plicata*

4'4"	**48'**	23'	106	1993	Lake Forest Park, 17402 44th Ave NE *RVP*
4'2"	**48'**	25'	104	1993	Lake Forest Park, 17402 44th Ave NE *RVP*

Japanese
Thuja standishii

2'10"	**33'**	29'	74	1995	Seattle, Washington Park Arboretum, 18N 7E *RVP*

Oriental
Thuja orientalis

4'5"	**46'**	23'	105	1995	Tacoma, Pt Defiance Park *RVP*

Pale Gold Siberian
Thuja occidentalis 'Wareana Lutescens'

2'4"	**22'**	9'	52	1993	Tacoma, Old Tacoma Cemetery *RGB*

Variegated Hiba
Thujopsis dolabrata 'Variegata'

5'6"	36'	23'	108	1995	Seattle, 1722 California Ave SW *RGB*
2'5" 1 of several stems	**52'**	29'	88	1987	Seattle, Leschi Park *ALJ*

Zebra
Thuja plicata 'Zebrina'

13'7"	65'	40'	238	1992	Renton, Greenwood Cemetery, 350 Monroe Ave NE *RGB*
12'2"	**76'**	51'	235	1993	Olympia, old Capitol building *ALJ, RVP*
10'7"	73'	**65'**	216	1989	Woodinville, Chateau Ste Michelle Winery *RVP*

ASH

Caucasian
Fraxinus angustifolia

8'5"	**93'**	51'	207	1990	Tacoma, Wright Park *RVP*
10'0"	70'	51'	203	1992	Orting, Washington Old Soldiers Home *RGB*

Circumference	Height	Crown Spread	AFA Points	Date Last Measured	Location and Nominators
European					*Fraxinus excelsior*
7'11"	87'	**81'**	**202**	1988	Walla Walla, Pioneer Park *RVP*
10'7"	59'	54'	**199**	1988	Mt Vernon, 1830 Beaver Marsh Rd *RVP*
6'8"	**93'**	59'	186	1990	Tacoma, Wright Park *RVP*
Flame					*Fraxinus angustifolia* 'Flame'
5'6"	**72'**	48'	**150**	1993	Seattle, 11351 35th Ave NE *ALJ*
5'9"	61'	**55'**	144	1993	Seattle, 10532 35th Ave NE *ALJ*
Flowering					*Fraxinus ornus*
7'0"	**69'**	43'	**164**	1992	Puyallup, Riverside Dr E and 78th St E *RGB, RVP*
7'9"	54'	**50'**	159	1993	Seattle, Volunteer Park *ALJ*
Green					*Fraxinus pennsylvanica*
12'7"	95'	**79'**	**266**	1995	Seattle, 610 36th Ave E *ALJ*
6'7"	**102'**	45'	192	1995	Seattle, 833 36th Ave E *RVP*
Weeping					*Fraxinus excelsior* 'Pendula'
5'5"	**38'**	**32'**	**111**	1993	Everett, 2131 Grand Ave *RGB*
White					*Fraxinus americana*
15'11"	90'	72'	**299**	1993	Rockport, 5341 N Cascades Hwy, E of Conrad Rd *RGB (below)*
8'3"	**101'**	48'	212	1992	Sedro Woolley, Northern State Multi Service Center *RGB, RVP*
13'10"	74'	**79'**	260	1990	Longview, S shore of Sacajawea Lake Park *RVP*

AZARA

Circumference	Height	Crown Spread	AFA Points	Date Last Measured	Location and Nominators
Boxleaf					*Azara microphylla*
3'0"	**21'**	**25'**	**63**	1995	Seattle, University of Washington, Friendship Grove *RGB, RVP*

The **White Ash** (left) is native to the eastern United States and is our nation's largest-growing species. Reaching optimum development on rich, bottomland soils, this giant along the Skagit River near Rockport has grown into Washington's largest planted ash.

The **American Beech** (above) is much less common in Washington than its European cousin, yet its elegant, richly colored leaves and slow growth give it a distinction all its own. This large example is in Olympia.

INTRODUCED TREES

Circumference	Height	Crown Spread	AFA Points	Date Last Measured	Location and Nominators
BALD CYPRESS					*Taxodium distichum*
8'0"	**96'**	31'	**200**	1993	Seattle, Green Lake Park, E tennis courts *ALJ*
8'6"	89'	37'	**200**	1993	Seattle, Green Lake Park near Corliss Ave N *RVP*
9'2"	71'	48'	**193**	1995	Tacoma, Pt Defiance Park *RVP*
Pond Cypress					*Taxodium ascendens*
3'10"	**63'**	19'	**114**	1989	Tacoma, Wright Park *RVP*
BASSWOOD (see also **LINDEN**, page 63)					
American					*Tilia americana*
9'10"	92'	**61'**	**225**	1990	Grays River, Swede Park *Bob Pyle*
8'2"	**94'**	**61'**	207	1988	Walla Walla, Pioneer Park *RVP*
White					*Tilia heterophylla*
11'5"	104'	68'	**258**	1993	Walla Walla, 221 Whitman *Shirley Muse, RGB, RVP*
8'11"	**110'**	63'	233	1990	Walla Walla, 343 Catherine *Shirley Muse, RVP*
8'4"	104'	**71'**	222	1988	Walla Walla, Pioneer Park *RVP*
BEECH					
American					*Fagus grandifolia*
10'8"	75'	**80'**	**223**	1993	Olympia, Sylvester Park *ALJ, RVP* (photo page 27)
5'4"	**85'**	59'	164	1987	Tacoma, Wright Park *RVP*
Columnar					*Fagus sylvatica* 'Dawyck'
4'5"	47'	21'	**105**	1993	Seattle, Lincoln High School *ALJ*
2'1"	**66'**	13'	**94**	1993	Seattle, Washington Park Arboretum, 41N 1E *ALJ*

This giant **European Beech** in Orting is reminiscent of the great beeches of its native homeland. Planted in 1875, this specimen is older than many of our other introduced trees, and among our most impressive.

The **Copper Beech** is the very popular, purple-leaved form of European Beech. We have several individuals over 5' in diameter, and this tree in an Everett cemetery is approaching 120' in height.

Circumference	Height	Crown Spread	AFA Points	Date Last Measured	Location and Nominators
Copper/Purple					*Fagus sylvatica* **f.** *purpurea*
16'0"	**119'**	81'	**331**	1993	Everett, Evergreen Cemetery *RVP* (photo page 28)
17'0"	91'	75'	**314**	1992	Blaine, 265 Boblett St *RGB*
15'8"	87'	**99'**	300	1988	Bellingham, 2231 Williams St *RVP*
European					*Fagus sylvatica*
19'7"	82'	**103'**	**343**	1995	Guemes Island, 698B Millett Rd *Piatt Bliss*
14'7"	104'	92'	**302**	1992	Orting, Washington Old Soldiers Home *RGB* (photo page 28)
6'3"	**121'**	53'	210	1992	Seattle, Washington Park Arboretum, 4S 8E *ALJ*
Fernleaf					*Fagus sylvatica* **'Asplenifolia'**
10'10"	**89'**	**76'**	**238**	1988	Walla Walla, Pioneer Park *RVP*
Purple Oakleaf					*Fagus sylvatica* **'Rohanii'**
2'11"	**58'**	**33'**	**101**	1993	Seattle, Washington Park Arboretum, 42N 1E *ALJ*
3'0"	44'	29'	87	1993	Richmond Beach, MSK Nursery, 20066 15th Ave NW *RGB, RVP*
Purple Tricolor					*Fagus sylvatica* **'Purpurea Tricolor'**
7'7"	**73'**	**52'**	**177**	1995	Brinnon, Whitney Gardens *RGB, RVP*
Weeping					*Fagus sylvatica* **'Pendula'**
5'9"	**68'**	**57'**	**151**	1993	Aberdeen, 3118 US Hwy 12 *RGB*

BIRCH

Circumference	Height	Crown Spread	AFA Points	Date Last Measured	Location and Nominators
Chinese Paper					*Betula albo-sinensis* **var.** *septentrionalis*
4'0"	35'	**30'**	**90**	1991	Mountlake Terrace, Holyrood Cemetery *RGB*
2'3"	**50'**	27'	84	1992	Seattle, Washington Park Arboretum, 25N 5W *RVP*
Columnar					*Betula pendula* **'Fastigiata'**
6'0"	**89'**	**37'**	**170**	1990	Redmond, Marymoor Park *RGB, RVP*
Cutleaf Weeping					*Betula pendula* **'Crispa'**
10'0"	73'	**72'**	**211**	1988	Yakima, 2702 W Yakima Ave *RVP* (photo below)
5'7"	**86'**	40'	163	1988	Wenatchee, Legion Park *RVP*
Downy					*Betula pubescens*
13'11"	67'	**78'**	**253**	1988	Aberdeen, 304 W Ninth St *RVP*
11'8"	**71'**	75'	230	1988	Aberdeen, 304 W Ninth St *RVP*

The small form and umbrella-shaped crown of the **Young's Weeping Birch** (above) lends a bizarre character to the gravestones at Calvary Cemetery in Seatttle.

European Birches seem to prefer the drier, harsher climate of eastern Washington to the damp, aphid-ridden conditions found west of the Cascades. The bright white trunk and vigorous constitution of the state's largest **Cutleaf Weeping Birch** (left) in Yakima are presented as evidence.

INTRODUCED TREES

Circumference	Height	Crown Spread	AFA Points	Date Last Measured	Location and Nominators
European White					*Betula pendula*
10'1"	**99'**	73'	**238**	1993	Snohomish, 330 Ave A *RVP*
11'8"	64'	**74'**	222	1987	Leavenworth, 12th & Front Sts *RVP*
Gray					*Betula populifolia*
5'5"	56'	44'	**132**	1989	Tacoma, Wright Park *RVP*
2'3"	**68'**	30'	102	1990	Bothell, Rhody Ridge Arboretum *RGB*
4'10"	45'	**48'**	115	1993	Tacoma, Proctor & N 30th *RVP*
Jacquemont					*Betula utilis* **var.** *jacquemontii*
3'7"	**75'**	31'	**126**	1988	Seattle, Washington Park Arboretum, M7 *ALJ*
Japanese White					*Betula platyphylla* **var.** *japonica*
6'0"	**91'**	55'	**177**	1989	Seattle, Washington Park Arboretum, Foster Island *RVP*
Paper					*Betula papyrifera*
10'0"	**98'**	71'	**236**	1993	Lake Forest Park, 17082 Brookside Blvd NE *RVP*
9'8"	91'	**72'**	225	1990	Tacoma, Wright Park *RVP*
Purple					*Betula pendula* **'Purpurea'**
2'6"	**57'**	29'	**94**	1993	Tacoma, 2117 Mountan View *KVP, RVP*

best of 2 stems

The attractive, flaking bark characteristic of the **River Birch** distingushes it from others of its clan. The state's largest examples of this graceful tree can be seen at the north parking area of Boeing's famed Museum of Flight, in Seattle.

The **Yellow Birch** is a large birch native to rich deciduous forests of the eastern US. Its shiny, brass colored bark makes it an ideal lawn specimen. This tree is at the Washington Old Soldiers Home in Orting, which is also home to several other record trees.

Circumference	Height	Crown Spread	AFA Points	Date Last Measured	Location and Nominators
River					*Betula nigra*
11'1"	68'	**98'**	**225**	1992	Tukwila, Museum of Flight *RGB* *(photo page 30)*
11'1"	68'	86'	**222**	1992	Tukwila, Museum of Flight *RGB*
7'11"	**86'**	82'	201	1987	Redmond, Marymoor Park *ALJ*
Sweet					*Betula lenta*
6'9"	55'	**64'**	**152**	1988	Seattle, Woodland Park Zoo *ALJ*
5'0"	**72'**	44'	143	1988	Seattle, Volunteer Park *ALJ*
Yellow					*Betula alleghaniensis*
11'6"	48'	74'	**204**	1992	Orting, Washington Old Soldiers Home *RGB* *(photo page 30)*
9'1"	68'	75'	**196**	1987	Seattle, University of Washington, Denny Hall *ALJ*
7'9"	79'	**79'**	192	1988	Seattle, University of Washington, Denny Hall *ALJ*
6'6"	**92'**	55'	184	1990	Lakewood, 26 Country Club Lane *KVP, RVP*
Young's Weeping					*Betula pendula* 'Youngii'
3'2"	23'	25'	**67**	1990	Seattle, Calvary Cemetery *RVP* *(photo page 29)*
2'6"	**27'**	**27'**	**64**	1990	Seattle, 815 NW 116th St *ALJ, RVP*
2'10"	22'	21'	61	1990	Seattle, 815 NW 116th St *ALJ, RVP*

BOX-ELDER

Circumference	Height	Crown Spread	AFA Points	Date Last Measured	Location and Nominators
					Acer negundo
12'5"	**69'**	**95'**	**242**	1990	Walla Walla, City Natural Area *Shirley Muse*
Variegated					*Acer negundo* 'Variegatum'
6'0"	**48'**	49'	132	1990	Woodland, Hulda Klager Lilac Gardens *RVP*
7'2"	32'	41'	128	1993	Selah, 1430 N Wenas Rd *Larry Rueter, RGB*
Yellow Variegated					*Acer negundo* 'Elegans'
7'4"	47'	48'	147	1992	Woodinville, Chateau St Michelle winery *RGB*
6'1"	40'	**51'**	126	1992	Monroe, 514 W Main St *RGB*

The bizarre shape of the National Champion **Northern Catalpa** adds a grotesque quality to the Phi Delta Omega House at Whitman College in Walla Walla. This is one of four National Champion trees found in the Walla Walla area.

The **Atlas Cedar**, native to the mountains of Morroco and Algeria, is one of our few African trees. It is also one of our most common, particularly in its blue form. Let this behemoth in Puyallup and the dimensions listed inform people how large these trees can really grow.

INTRODUCED TREES

Circumference	Height	Crown Spread	AFA Points	Date Last Measured	Location and Nominators

BOXWOOD
Buxus sempervirens

Circumference	Height	Crown Spread	AFA Points	Date Last Measured	Location and Nominators
4'7"	17'	12'	**75**	1993	Snohomish, GAR Cemetery *RGB*
1'0"	**28'**	**28'**	47	1993	Aberdeen, 3118 US Hwy 12 *RVP*
1 of several stems					

BUCKEYE   (see also **HORSECHESTNUT**, page 60)

California
Aesculus californica

Circumference	Height	Crown Spread	AFA Points	Date Last Measured	Location and Nominators
5'9"	38'	**45'**	118	1989	Seattle, Carl S English Gardens, Ballard Locks *ALJ*

Ohio
Aesculus glabra

7'4"	75'	43'	**174**	1993	Pullman, Reaney Park *RGB*
7'2"	**76'**	**44'**	173	1993	Pullman, Reaney Park *RGB*

Red
Aesculus pavia

2'5"	17'	**25'**	52	1992	Parkland, 127 127th St E *RGB, RVP*
1'10"	**21'**	**25'**	49	1993	Seattle, Carl S English Gardens, Ballard Locks *ALJ*

Yellow
Aesculus flava

10'3"	86'	56'	**223**	1988	Walla Walla, Pioneer Park *RVP*
9'3"	**95'**	**68'**	**223**	1988	Walla Walla, Pioneer Park *RVP*
11'11"	67'	45'	221	1992	Puyallup, Warner Karshner Elementary School *RGB, RVP*

In Western Washington, one of our more frequently planted blue-foliaged conifers is the **Blue Atlas Cedar**. I wonder if people realize just how large these trees can grow. This titan in Idylwild Park in Redmond attests to the tree's potential.

The **Deodar Cedar** is at once soft and graceful, yet gigantic, as illustrated by this Seattle specimen. This tree hails from the Himalayas and is an important timber tree there. Actually members of the pine family, the true cedars (genus *Cedrus*) include the Deodar, Atlas, and Lebanon.

Circumference	Height	Crown Spread	AFA Points	Date Last Measured	Location and Nominators
BUTTERNUT (see also **WALNUT**, pages 95-97)					*Juglans cinerea*
13'0"	71'	81'	**247**	1992	Clear Lake, 2193 Old Day Creek Rd *RGB*
10'11"	**77'**	59'	223	1987	Seattle, 9530 14th Ave NW *ALJ*
11'1"	73'	**85'**	227	1988	Sedro Woolley, 4th & Warner *RVP*

CATALPA

Ducloux					*Catalpa fargesii* **var.** *duclouxii*
3'1"	39'	24'	**82**	1993	Sumner, 136th Ave E and County Line Rd *RGB, RVP*
2'1"	**42'**	13'	70	1992	Seattle, 1900 Shenandoah Dr E *RGB*
Hybrid					*Catalpa* x *erubescens*
9'4"	**78'**	62'	**199**	1992	Tacoma, Wright Park near central restroom *RVP*
10'6"	46'	**65'**	188	1994	Seattle, Ballard Playground *ALJ*
Northern					*Catalpa speciosa*
22'0"	86'	**84'**	❖**371**	1993	Walla Walla, 715 Estrella, ΨΔΩ House *RVP* (photo page 31)
19'5"	**90'**	75'	342	1993	College Place, 421 Gose St *Shirley Muse*
Southern					*Catalpa bignonioides*
12'7"	37'	**70'**	**205**	1990	Woodland, 1423 Goerig St *RVP*
7'7"	**59'**	48'	162	1987	Tacoma, Wright Park *RVP*
Umbrella					*Catalpa bignonioides* **'Nana'**
7'6"	17'	25'	**113**	1990	Tacoma, N Prospect & Yakima *KVP, RVP*
5'2"	17'	**31'**	87	1990	Tacoma, Pt Defiance Park *RVP*
Yellow					*Catalpa ovata* **'Flavescens'**
10'4"	41'	**52'**	**178**	1992	Centralia, 216 W Walnut St *RGB*
3'10"	**52'**	35'	107	1993	Seattle, Washington Park Arboretum, 39N 3W *ALJ*

Everyone has held wood from the **Incense Cedar**, which is also called Pencil Cedar. Native to the mountains of Oregon and California, its bright green foliage and rich orange bark make this tree unmistakable. Our largest, pictured with little Ivan and the author, is in Point Defiance Park in Tacoma.

Japan's equivalent to our Giant Sequoia is the **Japanese Cedar**, or Sugi. Closely related and also capable of outsized proportions, the Japanese Cedar has foliage and cones that demand a second look. This specimen is also at Point Defiance Park, home to many record trees.

INTRODUCED TREES

Circumference	Height	Crown Spread	AFA Points	Date Last Measured	Location and Nominators

CEDAR (see also ARBORVITAE, page 26; and CYPRESS, page 45)

Atlantic White — *Chamaecyparis thyoides*

Circumference	Height	Crown Spread	AFA Points	Date Last Measured	Location and Nominators
2'10"	**36'**	25'	**76**	1988	Seattle, Cowen Park *ALJ*
3'6"	28'	19'	**75**	1990	Tacoma, University of Puget Sound *RVP*

Atlas — *Cedrus atlantica*

Circumference	Height	Crown Spread	AFA Points	Date Last Measured	Location and Nominators
14'10"	**107'**	**99'**	**310**	1992	Puyallup, Woodbine Cemetery *KVP, RVP (photo page 32)*
16'0"	79'	78'	291	1987	Seattle, 3101 W Laurelhurst Dr NE *ALJ*

Blue Atlas — *Cedrus atlantica* 'Glauca'

Circumference	Height	Crown Spread	AFA Points	Date Last Measured	Location and Nominators
16'4"	88'	**99'**	**309**	1995	Redmond, Idylwood Park *RGB (photo page 31)*
12'2"	**116'**	71'	280	1993	Olympia, McCleary Mansion *RVP*

Blue Sentinel — *Cedrus atlantica* 'Glauca Fastigiata'

Circumference	Height	Crown Spread	AFA Points	Date Last Measured	Location and Nominators
10'0"	84'	53'	**217**	1993	Seattle, 745 18th Ave E *RGB*
7'10"	**92'**	40'	196	1989	Everett, Forest Park, maintenance building *RVP*

Cock's Comb Japanese — *Cryptomeria japonica* 'Cristata'

Circumference	Height	Crown Spread	AFA Points	Date Last Measured	Location and Nominators
3'11"	**34'**	16'	**85**	1995	Montesano, 1417 Simpson *Neil Leonard, RGB*

Deodar — *Cedrus deodara*

Circumference	Height	Crown Spread	AFA Points	Date Last Measured	Location and Nominators
14'8"	84'	**83'**	**281**	1993	Longview, 2719 Garfield *RVP*
12'6"	102'	65'	**268**	1993	Seattle, 440 McGilvra Blvd E *RVP (photo page 32)*
9'11"	**114'**	52'	246	1988	Seattle, corner of E Howe St & Boyer Ave E *ALJ, RVP*

Eastern Red — *Juniperus virginiana*

Circumference	Height	Crown Spread	AFA Points	Date Last Measured	Location and Nominators
7'1"	**68'**	39'	**163**	1990	Yakima, 220 N 16th Ave *RVP*
7'3"	65'	33'	160	1993	Dayton, 1st Christian Church, Park & 3rd *Shirley Muse, RGB, RVP*

Golden Alaska — *Chamaecyparis nootkatensis* 'Lutea'

Circumference	Height	Crown Spread	AFA Points	Date Last Measured	Location and Nominators
12'7"	**60'**	39'	**221**	1995	Granite Falls, 1812 W Stanley St *RGB*

Golden Atlas — *Cedrus atlantica* 'Aurea Robusta'

Circumference	Height	Crown Spread	AFA Points	Date Last Measured	Location and Nominators
6'0"	49'	36'	**130**	1995	Mt Vernon, 1710 Old Hwy 99 S *RGB*
4'0"	**58'**	24'	112	1993	Fircrest, 514 Ramsdell *KVP, RVP*

Golden Deodar — *Cedrus deodara* 'Aurea'

Circumference	Height	Crown Spread	AFA Points	Date Last Measured	Location and Nominators
8'3"	**87'**	51'	**199**	1993	Vancouver, Evergreen Memorial Gardens *RGB*

Golden Japanese — *Cryptomeria japonica* 'Sekkan-Sugi'

Circumference	Height	Crown Spread	AFA Points	Date Last Measured	Location and Nominators
2'9"	**24'**	15'	**61**	1995	Tacoma, 7625 S Yakima *RGB*

Incense — *Calocedrus decurrens*

Circumference	Height	Crown Spread	AFA Points	Date Last Measured	Location and Nominators
15'10"	**104'**	49'	**306**	1995	Tacoma, Pt Defiance Park *ALJ, RVP (photo page 33)*

Japanese — *Cryptomeria japonica*

Circumference	Height	Crown Spread	AFA Points	Date Last Measured	Location and Nominators
15'5"	**101'**	55'	**300**	1995	Tacoma, Pt Defiance Park *ALJ, RVP (photo page 33)*

Japanese Plume — *Cryptomeria japonica* 'Elegans'

Circumference	Height	Crown Spread	AFA Points	Date Last Measured	Location and Nominators
6'10"	35'	18'	**121**	1993	Aberdeen, Fern Hill Cemetery *RGB, RVP*
4'0"	**47'**	17'	99	1995	Chico, 2641 NW Erlands Pt Rd *RGB*

Lebanese — *Cedrus libani*

Circumference	Height	Crown Spread	AFA Points	Date Last Measured	Location and Nominators
12'2"	**101'**	57'	**261**	1995	Snohomish, 10502 Lowell Rd *RGB*
13'6"	60'	**69'**	239	1990	Tacoma, Rust Mansion, N I St & N 10th St *KVP, RVP*

Northern White — *Thuja occidentalis*

Circumference	Height	Crown Spread	AFA Points	Date Last Measured	Location and Nominators
7'4"	56'	32'	**152**	1988	Walla Walla, Mountain View Cemetery *RVP*
5'10"	**68'**	23'	144	1988	Spokane, Pioneer Park *ALJ, RVP*

Port Orford — *Chamaecyparis lawsoniana*

Circumference	Height	Crown Spread	AFA Points	Date Last Measured	Location and Nominators
17'0"	110'	43'	**325**	1995	Olympia, Brewery Mansion *RVP*
13'10"	**113'**	39'	289	1995	Olympia, Brewery Mansion *RGB, RVP*

Robust Deodar — *Cedrus deodara* var. *robusta*

Circumference	Height	Crown Spread	AFA Points	Date Last Measured	Location and Nominators
5'10"	**30'**	40'	**110**	1992	Seattle, Franklin Place Park *RGB*

Sentinel — *Cedrus atlantica* f. *fastigiata*

Circumference	Height	Crown Spread	AFA Points	Date Last Measured	Location and Nominators
9'5"	75'	43'	**199**	1995	Seattle, 8018 18th Ave NW *RVP*

Variegated Alaska — *Chamaecyparis nootkatensis* 'Variegata'

Circumference	Height	Crown Spread	AFA Points	Date Last Measured	Location and Nominators
6'6"	51'	29'	**136**	1993	Tacoma, 2715 N Junett *RVP*
3'7"	**57'**	23'	106	1993	Seattle, 3617 E Columbia St *Mike Lee*

Variegated Atlantic White — *Chamaecyparis thyoides* 'Variegata'

Circumference	Height	Crown Spread	AFA Points	Date Last Measured	Location and Nominators
3'7"	**39'**	9'	**84**	1993	Milton, 101 Milton Way *RVP*

Circumference	Height	Crown Spread	AFA Points	Date Last Measured	Location and Nominators
Weeping Alaska					*Chamaecyparis nootkatensis* 'Pendula'
8'3"	**74'**	25'	**179**	1990	Tacoma, 7535 S Pine *KVP, RVP*
Weeping Blue Atlas					*Cedrus atlantica* 'Glauca Pendula'
4'1"	16'	**45'**	76	1992	Tukwila, Star Nursery, 13916 42nd Avve S *RGB*

CHERRY (see also **Japanese Winter-Flowering Cherries**, page 37; and **Japanese 'Sato' Cherries**, pages 38-39)

Circumference	Height	Crown Spread	AFA Points	Date Last Measured	Location and Nominators
Birchbark					*Prunus serrula*
5'0"	39'	40'	**109**	1993	Seattle, Ayre, The Highlands *RGB*
4'10"	37'	**41'**	105	1992	Milton, 24th St E *Dick North*
3'7"	**46'**	35'	98	1992	Lakewood, Lakewold Gardens *KVP, RVP*
Bird					*Prunus padus*
5'8"	43'	33'	**119**	1993	Tacoma, 3502 Oas Dr W *KVP, RVP*
3'8"	**54'**	**44'**	109	1993	Stanwood, Old Hwy 99 N & 36th Ave E *RGB*
Black					*Prunus serotina*
7'4"	73'	**70'**	178	1987	Seattle, Woodland Park Zoo *ALJ*
8'10"	50'	55'	**170**	1987	Seattle, Mt Pleasant Cemetery *ALJ*
4'10"	**88'**	41'	156	1993	Newhalem, at edge of slope near monument *RGB, RVP*
Choke					*Prunus virginiana*
3'6"	**48'**	**37'**	99	1993	Richmond Beach, MSK Nrsry, 20066 15th Ave NW *M Kruckeberg*
Common					*Prunus avium*
13'8"	84'	65'	**264**	1992	Preston, 29700 SE High Pt Rd *RGB*
14'6"	64'	**70'**	255	1995	Montesano, 11th St, N of Marcy St W *Neil Leonard, RGB*
5'6"	**100'**	28'	173	1993	Seattle, Interlaken Park *ALJ*

Anyone who has not witnessed the University of Washington Quad when the **Yoshino Cherries** are in bloom has missed one of our greatest floral displays. When the weather cooperates in early spring, the cameras outnumber the blossoms as folks try to capture the spectacle on film.

The **Daybreak Cherry** is a cultivar of Yoshino with larger flowers that are slightly darker pink. Its denser displays make this 'new and improved' Yoshino more frequently planted. Old specimens are rare, however. Our largest (with Iris for scale) may be seen in the Tacoma Country Club parking lot.

INTRODUCED TREES

Circumference	Height	Crown Spread	AFA Points	Date Last Measured	Location and Nominators
Daybreak					*Prunus* x *yedoensis* 'Akebono'
6'5"	**28'**	47'	**117**	1992	Seattle, 2061 Interlaken Pl E, on Peach Ct *ALJ*
6'8"	21'	**49'**	113	1993	Lakewood, Tacoma Country Club *KVP, RVP (photo page 35)*
Double-Flowered Mazzard					*Prunus avium* 'Plena'
6'6"	**52'**	49'	**142**	1993	Seattle, Washington Park Arboretum, 10N 0 *RVP*
6'6"	51'	**53'**	**142**	1993	Seattle, Evergreen Cemetery *KVP, RVP*
Fuji					*Prunus incisa*
2'10"	**25'**	**34'**	**67**	1993	Seattle, Washington Park Arboretum, 31N 5E *ALJ*
Goldbark					*Prunus maackii*
3'8"	**40'**	**39'**	**94**	1990	Spokane, Finch Arboretum *ALJ, RVP*
Hillier					*Prunus* 'Hillieri'
8'5"	**30'**	**56'**	**145**	1993	Seattle, 10709 39th Ave NE *RGB*
5'0"	**34'**	39'	104	1995	Seattle, Washington Park Arboretum, 30N 1W *KVP, RVP*
Japanese Hill					*Prunus jamasakura*
10'5"	48'	**59'**	**188**	1993	Vancouver, Park Hill Cemetery *RGB*
6'0"	**58'**	55'	144	1993	Seattle, Washington Park Arboretum, 26N 2W *ALJ*
Korean Hill					*Prunus verecunda*
8'9"	**53'**	52'	**171**	1993	Newhalem, E end of central lawn area *RGB, ALJ*
7'11"	**60'**	**55'**	169	1993	Newhalem, N of road in lawn area *RGB, ALJ*
Miyama					*Prunus maximowiczii*
6'1"	**50'**	**40'**	**133**	1993	Newhalem, E central part of main lawn *RGB, ALJ*
Naden					*Prunus* x *sieboldii*
8'0"	19'	31'	**123**	1993	Seattle, 712 12th Ave E *ALJ*
4'6"	**24'**	**40'**	88	1993	Seattle, 1002 37th Ave E *ALJ, RVP*
Oshima					*Prunus speciosa*
7'4"	**35'**	40'	**133**	1990	Seattle, Williams Place, 15th Ave E & E John St *ALJ*
5'3"	30'	**51'**	106	1995	Bremerton, Naval Shipyard, bldg 644 *ALJ*

The cherry shown is a form of the Sato Zakura Cherries known as **Hokusai**. It may be found in Point Defiance Park in Tacoma. It is considered by some (including myself) as the finest flowering cherry tree in Washington. Not only is this venerable specimen one of our largest flowering cherries (it is nearly 60 feet wide!), but as of this writing it is unique - no other old specimens are known in the region.

Japanese Winter-Flowering Cherries

Higan Cherries, *Prunus* x *subhirtella*, as they are commonly known in the United States, are a group of early-flowering cherries from Japan. The x in the scientific name refers to hybridity, the parents being *P. incisa* and *P. pendula*. *P. incisa* is a small, delicate tree whereas *P. pendula* is a vigorous, robust tree. According to the Japanese, who developed these hybrids, the classification needs to be updated. The trees listed below are now put into two categories: (1) *P.* x *subhirtella* and: (2) *P. pendula*. In addition, two hybrids are listed; 'Accolade', which is *P.* x *subhirtella* crossed with *P. sargentii*, and 'Pandora', which is *P.* x *subhirtella* 'Rosea' crossed with *P.* x *yedoensis*.

Taxon	Circumference	Height	Crown Spread	AFA Points	Date Last Measured	Location and Nominators
Prunus x **subhirtella**						
'Autumnalis Rosea'	5'6"	29'	**40'**	**105**	1990	Seattle, 6026 41st Ave NE *ALJ*
	3'6"	**36'**	29'	85	1990	Seattle, University of Washington, Benson Hall *ALJ*
'Whitcomb'	**8'1"**	36'	**55'**	**145**	1995	Monroe, 302 S Blakely St *RGB, RVP (photo below)*
	8'1"	27'	51'	137	1993	Seattle, Washelli Cemetery *RVP*
	3'6"	**39'**	36'	90	1990	Seattle, 1018 36th Ave E *ALJ, RVP*
Prunus pendula						
	8'10"	**48'**	51'	**167**	1993	Walla Walla, 210 Marcus St *Shirley Muse, RGB, RVP*
	8'5"	33'	**54'**	147	1992	Kent, 22821 104th Ave SE *RGB*
var. *ascendens*	6'9"	52'	40'	**143**	1992	Woodinville, Chateau Ste Michelle Winery *RGB*
	6'10"	34'	**53'**	129	1993	Lakewood, Tacoma Cntry Club *KVP, RVP*
	4'8"	**57'**	47'	125	1993	Woodinville, Chateau Ste Michelle Winery *RGB*
'Pendula Plena Rosea'	5'9"	**18'**	**31'**	95	1993	Puyallup, 5821 108th Ave Ct E *KVP, RVP*
'Pendula Rubra'	**8'5"**	**26'**	**40'**	137	1993	Camas, 1511 NE 3rd Place *RVP*
'Stellata'	**5'3"**	**25'**	**33'**	96	1992	Seattle, Washington Park Arboretum, 20N 2W *RVP*
Prunus Hybrids						
'Accolade'	2'9"	23'	**47'**	68	1993	Seattle, Washington Park Arboretum, 14N 3E *RVP*
	2'6"	**26'**	39'	66	1993	Seattle, Washington Park Arboretum, 14N 3E *RVP*
	2'7"	23'	33'	65	1993	Tacoma, 511 N Sheridan *KVP, RVP*
'Pandora'	**4'0"**	**51'**	**41'**	109	1992	Lakewood, Lakewold Gardens *KVP, RVP*

Iris is seen here admiring the state's largest **Whitcomb Cherry**, in Monroe. The Whitcomb is a Northwest original, as the cultivar was named by David Whitcomb in 1925 at his estate in Edmonds. More vigorous than most Higans, the dark pink flowers brighten up February throughout much of the Puget Sound region.

INTRODUCED TREES

Japanese 'Sato' Cherries

Sato, or Sato Zakura, is a Japanese term for purely ornamental cherries. Commonly classified as cultivars of **Prunus serrulata**, this Japanese species is not the parent to the trees commonly associated with it. These cultivars result from a complex of hybrids and backcrosses. Most have blood from one or more of the native mountain cherries of Japan: *Prunus lannesiana, P. speciosa, P. jamasakura* or *P. verecunda*. Their variety in parentage is surpassed only by their variety of bloom. Flowering is spread out over two months in spring, from 'Shirotae' in mid-March to 'Shogetsu' in late May. Flowers range from the large single white blooms of 'Tai Haku' 2½ inches across, to the delicate pink petals of 'Fugenzo', which number 30 or more per flower. For showiness at time of flowering their popularity is unparalleled, and their dimensions indicate great vigor in our climate.

Cultivar	Circumference	Height	Crown Spread	AFA Points	Date Last Measured	Location and Nominators
'Amanogawa'	4'7"	38'	32'	101	1993	Seattle, Washelli Cemetery, NE of mausoleum *ALJ*
'Choshu-hizakura'	3'10"	30'	36'	85	1993	Seattle, University of Washington, Red Square *ALJ, RVP*
'Fugenzo'	4'9"	21'	31'	86	1993	Seattle, Washington Park Arboretum, 41N 2E *RVP*
'Hisakura'	4'4"	25'	45'	88	1995	Seattle, 116 NW 60th St *RVP*
'Hokusai'	6'11"	24'	57'	121	1990	Tacoma, Point Defiance Park *RVP (photo page 36)*
'Horinji'	4'3"	56'	55'	121	1993	Seattle, Washington Park Arboretum, 10N 0 *ALJ*
	6'1"	34'	42'	117	1993	Seattle, Hiawatha Playground, wading pond *ALJ*
'Kwanzan'	9'4"	39'	55'	165	1992	Sumner, 1302 Main, tree on Sumner Ave *RVP (photo opp)*
	5'1"	50'	41'	121	1993	Sultan, 106 4th Ave, back yard *RGB, RVP*
	6'2"	48'	60'	137	1995	Brinnon, Whitney Gardens *KVP, RGB, RVP*
'Mikuruma-gaeshi'	4'4"	21'	39'	83	1993	Seattle, Volunteer Park, N of bandshell *ALJ*
'Ojochin'	4'5"	25'	28'	85	1993	Fircrest, 216 Columbia Ave *RGB, RVP*
'Pink Perfection'	3'5"	20'	35'	70	1993	Tacoma, 1301 N Proctor *KVP*
'Shirofugen'	9'1"	27'	47'	148	1993	Tacoma, 4163 Alki St *RVP*
	7'6"	38'	49'	140	1993	Vancouver, Park Hill Cemetery *RGB*
'Shirotae'/'Mt Fuji'	7'8"	21'	43'	124	1987	Seattle, Woodland Park Zoo *ALJ*
	5'10"	37'	51'	120	1990	Seattle, Washington Park Arboretum, 33N 2W *RVP*
'Shogetsu'	5'4"	13'	29'	84	1992	Tacoma, Mountain View Cemetery *RVP*
	2'6"	17'	40'	57	1993	Rockport, 5 mi E on Hwy 20 *RGB, RVP*
'Tai Haku'	3'7"	29'	37'	81	1992	Lakewood, Lakewold Gardens near main bldg *RVP*
'Taoyoma'	3'0"	18'	39'	64	1993	Tacoma, 1221 N Proctor *KVP*
'Temari'	4'7"	22'	39'	87	1993	Seattle, 31st Ave NE at NE 82nd St *ALJ, RVP*
	3'6"	26'	31'	76	1993	Seattle, University of Washington, Hansee Hall *ALJ*
'Ukon'	5'6"	33'	49'	111	1995	Tacoma, S 7th & Tacoma Ave *KVP, RVP*
	4'9"	42'	36'	108	1993	Tukwila, Towne and Country Motel *RGB*
	6'1"	25'	29'	105	1995	Tacoma, S K St & S 8th St *KVP, RVP*

—*Hokusai*—

An old and rare cultivar, Hokusai was introduced into Europe in 1866 and named for a famous 19th-century Japanese artist. The tree in Point Defiance Park may be the largest of this type in North America.

—*Shogetsu*—

The last of the locally familiar Sato cherries to bloom, this is one of the best. It has double, pure white flowers and slowly grows into a small tree of great distinction.

—*Tai Haku*—

The Japanese name means 'Great White Cherry', as it has the largest flowers of any cherry. It was lost to cultivation, before being rediscovered by Collingwood Ingram in England. Its vigorous constitution and robust form allow it to grow into a beautiful, broad-shaped tree.

Far and away the most vigorous (and perhaps the gaudiest as well), the **Kwanzan Cherry** is very common in parks and gardens of western Washington. This example in Sumner is our largest.

The **Ukon Cherry** is one of the most distinct of the Sato cherries. Its pleasant yellow-green flowers distinguish it from the pink or white varieties. Ivan is up in the branches of this University of Washington champion, which has been recently removed.

INTRODUCED TREES

	Circumference	Height	Crown Spread	AFA Points	Date Last Measured	Location and Nominators
Sargent						*Prunus sargentii*
	7'1"	36'	**43'**	**132**	1992	Seattle, Washington Park Arboretum, 28N 2W *ALJ, RVP*
	3'11"	**45'**	24'	98	1993	Puyallup, WSU Experiment Station *RGB, RVP*
Spaeth						*Prunus padus* 'Spaethii'
	4'2"	**40'**	**35'**	99	1995	Seattle, 1005 S Southern St *ALJ*
St Lucie						*Prunus mahaleb*
	6'6"	36'	**47'**	❖**126**	1993	Lake Stevens, 8420 123rd Ave NE *RGB*
	6'9"	27'	40'	**118**	1993	Laurel, 6511 Guide Meridian Rd *RGB*
	3'6"	**37'**	43'	90	1992	Woodinville, Chateau Ste Michelle Winery *RGB*
Yoshino						*Prunus* x *yedoensis*
	9'6" below forking	41'	**65'**	171	1993	Seattle, University of Washington, Quad *RVP* *(photo page 35)*
	9'4"	46'	48'	**170**	1993	Newhalem, in lawn at W end of town *RGB*
	3'5"	**52'**	35'	102	1987	Seattle, Lincoln Park *ALJ*

CHESTNUT

	Circumference	Height	Crown Spread	AFA Points	Date Last Measured	Location and Nominators
American						*Castanea dentata*
	19'7"	**106'**	101'	❖**366**	1993	Cicero, mile post 30 on St Hwy 530 *RVP* *(photo below)*
	20'7"	87'	**109'**	❖**361**	1993	Carson, mile post 1.4 on Metzger Rd *RGB, RVP*

This **American Chestnut** is a National Champion and may be found in a field at Cicero, not far from the waters of the N Fork Stillaguamish River. The great chestnut forests that once cloaked the Appalachian Mountains were eliminated by chestnut blight during the 1920's and 30's. Most of the remaining specimens are found in western states – brought out by pioneers. The blight has not reached these trees and Washington has the two largest American Chestnuts known in the world. Our other record chestnut is at the town of Carson, located in the heart of the Columbia Gorge. In a species famous for rapid growth, Washington's two record trees have grown to over six feet in diameter in around 100 years.

American Chestnut

Circumference	Height	Crown Spread	AFA Points	Date Last Measured	Location and Nominators
Chinese					*Castanea mollissima*
5'7"	54'	**69'**	**138**	1988	Seattle, Lincoln Park *ALJ*
6'1"	44'	61'	**132**	1992	Parkland, 316 140th St S *KVP, RVP*
4'3"	**64'**	51'	128	1988	Seattle, Lincoln Park *ALJ*
Spanish					*Castanea sativa*
20'8"	88'	71'	**354**	1988	Olympia, Chestnut Hill Dr *Michael Dolan*
21'5"	77'	76'	**352**	1992	Vashon Island, behind 17205 99th St SW *Mike Lee, RGB, RVP*
13'5"	**91'**	65'	268	1987	Tacoma, Wright Park *RVP*
16'9"	82'	**96'**	307	1988	Tacoma, N 27th & Alder *Tom Davis*
Sweet American (Hybrid)					(?) *Castanea* x *blaringhemii*
16'0"	**90'**	**104'**	**308**	1993	Tumwater, Masonic Cemetery *RGB*

CHINA FIR

					Cunninghamia lanceolata
10'10"	82'	32'	**220**	1995	Kent, S of James St & Russel Rd in Riverbend Golf complex *RGB*
6'1"	**91'**	28'	171	1993	Vashon Island, 16820 McIntyre Rd SW *RGB*
Blue					*Cunninghamia lanceolata* 'Glauca'
4'7"	**51'**	28'	**113**	1995	Seattle, 14048 Palatine Ave N *RVP*

CRABAPPLE (see also **Purple Crabapples**, page 43)

Aspiration					*Malus baccata* 'Aspiration'
4'6"	**42'**	**37'**	**105**	1992	Seattle, University of Washington, fountain area *ALJ*
Bechtel					*Malus ioensis* 'Plena'
2'11"	**29'**	**27'**	**71**	1993	Tacoma, 407 N E St *KVP, RVP*
Blanche Ames					*Malus* 'Blanche Ames'
6'3"	**30'**	**41'**	**115**	1992	Seattle, University of Washington, fountain area *ALJ*
Cherry					*Malus* x *robusta*
3'5"	**35'**	**32'**	**84**	1992	Seattle, Carl S English Gardens, Ballard Locks *ALJ*
Chinese Flowering					*Malus spectabilis* 'Plena'
3'2"	25'	29'	**70**	1992	Seattle, University of Washington, fountain area *ALJ*
2'6"	**33'**	28'	**70**	1992	Seattle, Washington Park Arboretum, 3S 9E *ALJ*
Cutleaf					*Malus toringoides*
4'9"	26'	**32'**	**92**	1993	Spokane, Finch Arboretum *ALJ*
2'7"	**31'**	30'	69	1992	Seattle, Washington Park Arboretum, 11N 2W *ALJ*
Dawson					*Malus* x *dawsoniana*
9'3"	49'	**52'**	**173**	1992	Issaquah, 414 Front St N *RGB, RVP (photo page 42)*
4'5"	**80'**	45'	144	1993	Seattle, Discovery Park *ALJ*
Dolgo					*Malus* 'Dolgo'
4'0"	31'	**39'**	**89**	1993	Seattle, 3768 University Way NE *ALJ*
2'10"	**37'**	35'	80	1992	Seattle, Washington Park Arboretum, 32N 4E *ALJ*
Dorothea					*Malus* 'Dorothea'
3'4"	**17'**	**27'**	**64**	1993	Tacoma, 1116 Skyline Dr *KVP, RVP*
Hillier					*Malus* x *scheideckeri* 'Hillieri'
4'0"	**29'**	**35'**	**86**	1993	Edmonds, 22006 76th Ave W *RVP*
Himalayan					*Malus baccata* var. *himalaica*
5'0"	**38'**	**41'**	**108**	1992	Seattle, Carl S English Gardens, Ballard Locks *ALJ*
Hupeh					*Malus hupehensis*
4'3"	25'	**40'**	**86**	1992	Seattle, University of Washington, stadium (SE) *ALJ*
3'5"	**36'**	36'	**86**	1992	Seattle, Washington Park Arboretum, 30N 4E *ALJ*
Japanese Flowering					*Malus* x *floribunda*
4'8"	**33'**	44'	**100**	1993	Seattle, 1254 Federal Ave E *RVP*
4'9"	21'	**47'**	90	1992	Tacoma, Point Defiance Park, Japanese Garden *RVP*

INTRODUCED TREES

Circumference	Height	Crown Spread	AFA Points	Date Last Measured	Location and Nominators
Kaido					*Malus* x *micromalus*
3'11"	27'	37'	**83**	1993	Tacoma, 502 N Yakima *KVP, RVP*
3'1"	**37'**	**38'**	**83**	1993	Seattle, Lincoln Park *ALJ, RVP*
3'5"	30'	35'	80	1993	Tacoma, 619 N Yakima *KVP, RVP*
Katherine					*Malus* 'Katherine'
1'10"	21'	**24'**	**49**	1994	Seattle, Washington Park Arboretum, near Madison St *ALJ*
1'7"	**24'**	21'	48	1993	Puyallup, WSU Experiment Station *RVP*
Klehm's Improved Bechtel					*Malus coronaria* 'Klehm's Improved Bechtel'
2'8"	**25'**	27'	**64**	1993	Seattle, Broadview Elementary School, Greenwood Ave N *RVP*
2'9"	22'	**28'**	62	1995	Yakima, Yakima Area Arboretum *RVP*
Manchurian					*Malus baccata* var. *mandshurica*
4'2"	**38'**	**41'**	**98**	1992	Seattle, University of Washington, fountain area *ALJ*
Parkman					*Malus halliana* 'Parkmanii'
2'2"	**22'**	**26'**	**54**	1993	Seattle, 1002 36th Ave E *RVP*
Pillar					*Malus tschonoskii*
2'10"	**35'**	**25'**	**75**	1995	Seattle, 725 S Rose St *RGB, RVP*
Prairie					*Malus ioensis*
2'4"	**21'**	**27'**	**56**	1993	Seattle, Seattle Center, near main entrance *ALJ*
Prince George's					*Malus* 'Prince Georges'
1'9"	**19'**	**23'**	**46**	1993	Lynnwood, Rhody Ridge Arboretum *RGB*
Red Jade					*Malus* 'Red Jade'
2'8"	**14'**	**29'**	**53**	1993	Seattle, Washington Park Arboretum, 2S 6E *ALJ, RVP*
Sargent					*Malus sargentii*
1'11"	**15'**	**31'**	**46**	1993	Seattle, Woodland Park Zoo, Rose Garden *ALJ*
Scheidecker					*Malus* x *scheideckeri*
3'2"	**21'**	**24'**	**65**	1992	Seattle, University of Washington, fountain area *ALJ*

The lawns near Drumheller Fountain at the University of Washington contain the best collection of crabapples in the state. Planted during the 1950's, the groves contain 9 State Champion trees. One of the most attractive is **Blanche Ames**, a gracefully weeping, white-flowered tree, shown here.

The **Dawson Crabapple** is a hybrid between the native Oregon Crabapple (*Malus fusca*) and the Common Orchard Apple (*Malus* x *domestica*). Gaining vigor from both parents, this hybrid is the largest crabapple in the state. Iris is shown in our largest Dawson Crab, a 3-foot diameter tree in Issaquah.

Purple Crabapples

Purple Crabs, also known as the Rosy Bloom Crabs, are both a delight and an eyesore in Northwest parks and gardens. When in bloom, some of these may be regarded as the most beautiful of all flowering trees. But for fifty weeks of the year, some of these same trees may be considered among the ugliest. The Purple Crabs are a group of hybrids and cultivars that share parentage to a small tree found in central Asia: the Redvein Crab, *Malus sieversii* 'Niedzwetzkyana'. This tree is nearly extinct in the commercial trade. Among the earlier hybrids was *Malus x purpurea*, a tree common only in our older parks. A great multitude of cultivars have largely replaced these in the nurseries, each having its own variation on the rosy flower theme. Most are highly susceptible to scab diseases, particularly when grown west of the Cascade Mountains. Regardless of their drawbacks as a group, their popularity remains high; as for two weeks in spring they are a truly spectacular sight.

Cultivar	Circumference	Height	Crown Spread	AFA Points	Date Last Measured	Location and Nominators
Redvein						*Malus sieversii* 'Niedzwetzkyana'
	4'5" below branching	28'	47'	93	1993	Woodland, Hulda Klager Lilac Garden *RGB, RVP*
Malus x *purpurea*	4'10"	41'	42'	109	1987	Seattle, University of Washington, Denny Hall *ALJ*
Cultivars						
'Almey'	2'8"	20'	31'	60	1993	Lynnwood, Rhody Ridge Arboretum *RGB*
'Echtermeyer'	2'11"	12'	24'	53	1992	Spokane, Finch Arboretum *RGB, ALJ*
	1'10"	24'	19'	51	1993	Walla Walla, Pioneer Park *RGB, RVP*
'Eleyi'	5'7"	31'	45'	109	1993	Longview, Huntington Park *RVP*
'Hopa'	4'7"	35'	47'	102	1992	Tacoma, Wright Park *RVP*
'Lemoinei'	3'1"	27'	33'	72	1995	Yakima, Yakima Area Arboretum *RVP*
'Liset'	3'7"	27'	35'	79	1993	Seattle, Seattle University *ALJ*
'Pink Beauty'	4'9"	24'	46'	92	1992	Seattle, Univ of Wash, Communication Bldg *ALJ, RVP*
'Profusion'	3'6"	35'	40'	87	1992	Seattle, Washington Park Arboretum, 32N 6E *ALJ*
'Radiant'	2'11"	29'	29'	71	1993	Seattle, Washington Park Arboretum, 16N 3W *RVP*
'Royalty'	3'4"	27'	40'	77	1993	Ferndale, W Axton Rd & Northwest Dr *RGB, RVP*

Pink Beauty is the name of this tree (left) at the UW in Seattle. Its name is appropriate for, in bloom, this variety is spectacular.

One of the most common of the Purple Crabs, **Hopa** is characterized by large, light purplish-pink flowers. The example shown (right) is in Tacoma's Wright Park.

INTRODUCED TREES

Circumference	Height	Crown Spread	AFA Points	Date Last Measured	Location and Nominators
Siberian					*Malus baccata*
6'3"	**41'**	**47'**	**128**	1992	Seattle, University of Washington, fountain area *ALJ*
6'6"	30'	41'	118	1992	Seattle, University of Washington, fountain area *ALJ*
Siberian Crab Hybrid					*Malus* x *adstringens*
5'6"	**48'**	**47'**	**126**	1992	Seattle, University of Washington, fountain area *ALJ*
Sweet					*Malus coronaria*
2'6"	**57'**	24'	93	1990	Seattle, Washington Park Arboretum, 32N 1E *ALJ*
3'9"	37'	**30'**	89	1992	Seattle, University of Washington, fountain area *ALJ*
Toringo					*Malus sieboldii* var. *arborescens*
5'1"	32'	33'	**101**	1992	Seattle, Lock Vista Apartments, 3025 NW Market St *ALJ*
4'0"	**34'**	**38'**	91	1992	Seattle, Lock Vista Apartments, 3025 NW Market St *ALJ*
Van Eseltine					*Malus* 'Van Eseltine'
4'6"	27'	**39'**	91	1993	Puyallup, 10918 57th St E *RGB, RVP*
Yellow Autumn					*Malus* x *sublobata*
3'9"	**30'**	**30'**	82	1992	Spokane, Finch Arboretum *ALJ*
Zumi					*Malus* x *zumi*
2'2"	**19'**	**29'**	52	1993	Seattle, Washington Park Arboretum, 11N 3W *RVP*

CRAPE-MYRTLE
Lagerstroemia indica

Circumference	Height	Crown Spread	AFA Points	Date Last Measured	Location and Nominators
1'5"	**32'**	19'	54	1990	Seattle, University of Washington, Herb Garden *ALJ*

The true cypresses fall under the scientific name *Cupressus.* Famous examples of this group are the tall narrow Italian Cypresses commonly seen in the Mediterranean region and the Monterey Cypress (*right.*). Many Cypress trees are native to California, including the **Arizona Cypress.** Located in Carson, the species shown is one of many to 'escape' the Forest Service Arboretum in Wind River.

The **Monterey Cypress** is native to a few small locales in and around Monterey, California. Generally small, windswept trees often featured on postcards, they grow to great size when planted in parks and gardens. It is one of the largest trees in Britain, and a California specimen is nearly 14' in diameter. Most places in Washington are too cold for this tree, but Port Townsend has managed to grow this 7' diameter giant.

Circumference	Height	Crown Spread	AFA Points	Date Last Measured	Location and Nominators

CYPRESS (see also ARBORVITAE, page 26; CEDAR, page 34; Lawson Cypress, pages 46-47; and Sawara Cypress, page 48)

Arizona
Cupressus arizonica

9'6"	68'	35'	**191**	1990	Carson, Carson Grange *RVP*

Arizona Smooth
Cupressus arizonica var. *glabra*

6'0"	57'	23'	**135**	1987	Seattle, Acacia Cemetery *ALJ*
4'11"	**62'**	20'	126	1988	Seattle, NE corner of E Madison & E Lee Sts *ALJ*

Castlewellan
X *Cupressocyparis leylandii* 'Castlewellan'

2'4"	40'	11'	**71**	1995	Bellevue, 3120 125th Ave SE *RGB*

one of many stems

Columnar Italian
Cupressus sempervirens 'Stricta'

2'3"	42'	6'	**70**	1988	Seattle, Harvard Ave E & E Gwinn Pl *ALJ, RVP*

Contorted Leyland
X *Cupressocyparis leylandii* 'Contorta'

4'10"	33'	35'	**100**	1995	Olympia, Puget St & Oak Ave *RGB*

Golden Hinoki
Chamaecyparis obtusa 'Crippsii'

4'0"	41'	24'	**83**	1995	Sedro Woolley, 2058 Cook Rd *RGB*

largest of two

Hinoki
Chamaecyparis obtusa

4'11"	53'	26'	**118**	1988	Seattle, 1160 20th Ave E *ALJ*

Italian
Cupressus sempervirens

3'3"	33'	15'	**76**	1993	Vashon Island, 17722 Vashon Hwy SW *Mike Lee*
1'9"	**37'**	11'	61	1993	Seattle, University of Washington, Anderson Hall *RVP*

Leyland
X *Cupressocyparis leylandii* 'Leighton Green'

7'10"	70'	33'	**172**	1990	Lakewood, Lakewold Gardens *KVP, RVP*
6'6"	**80'**	40'	168	1993	Seattle, Washington Park Arboretum, 36N 4W *RVP*
7'4"	69'	37'	166	1989	Seattle, Washington Park Arboretum, 1S 6E *ALJ*

Mendocino
Cupressus goveniana var. *pygmaea*

5'10"	**75'**	32'	**153**	1988	Seattle, Washington Park Arboretum, 12N 9E *RVP*

Monterey
Cupressus macrocarpa

23'7"	77'	64'	**378**	1995	Port Townsend, 430 Lawrence *RVP*
22'4"	79'	**79'**	367	1995	Port Townsend, 1310 Clay St *RVP*
21'8"	84'	77'	363	1988	Fox Island, near bridge *RVP*
16'8"	**100'**	59'	315	1995	Oysterville *RVP*

Piute
Cupressus nevadensis

5'9"	56'	25'	**131**	1993	Seattle, Washington Park Arboretum, 36N 5W *RVP*
4'2"	**69'**	24'	125	1993	Seattle, Washington Park Arboretum, 36N 5W *RVP*

Slender Hinoki
Chamaecyparis obtusa 'Gracilis'

2'2"	**26'**	21'	**57**	1993	Tacoma, Clover Park Voc Tech, E of Arboretum *RGB, RVP*

Variegated Leyland
X *Cupressocyparis leylandii* 'Gold Spot'

2'11"	**31'**	20'	**71**	1995	Seattle, Washington Park Arboretum, 43N 5W *KVP, RVP*

DOGWOOD

Cornelian-cherry
Cornus mas

3'3"	**39'**	**33'**	**89**	1995	Olympia, Brewery Mansion *RGB, RVP*
3'7"	33'	27'	83	1990	Edmonds, 9302 192nd St SW *RGB*

Eastern
Cornus florida

2'1"	**61'**	20'	**91**	1989	Seattle, Washington Park Arboretum, 19N 2E *ALJ*
3'10"	38'	**35'**	83	1990	Longview, Huntington Park *RVP*

Eddie's White Wonder
Cornus 'Eddie's White Wonder'

1'9"	**33'**	21'	**59**	1993	Bothell, Rhody Ridge Arboretum *RGB, RVP*

Kousa
Cornus kousa

5'7"	28'	**41'**	**105**	1987	Seattle, Laurelhurst Playfield *ALJ*
2'6"	**41'**	25'	77	1990	Seattle, 1018 36th Ave E *ALJ*

INTRODUCED TREES

Lawson Cypress

Lawson Cypresses were named for Peter Lawson of Edinburgh, Scotland, who imported the first seed in 1854. All Lawson Cypresses are cultivated varieties of the Port Orford Cedar, *Chamaecyparis lawsoniana*, a conifer native to the Siskiyou Mountains in southwestern Oregon and northwestern California. An interesting species from many viewpoints, Port Orford Cedar has a very small natural geographic range, yet within that range it grows on a wide variety of elevations and habitats. In the 150 years since its discovery, over 200 cultivars have been developed - more than any other conifer! The oldest cultivar is 'Erecta Viridis', which arose from the original seed batch of 1854. Of the 200+ cultivars, 30-40 attain tree-size dimensions. Although most of these were developed in Great Britain, all are adapted to the climate of western Washington. Lawson Cypress cultivars have proven to be one of the most difficult groups treated in this book to identify. A handful of cultivars accounts for 90% of the trees normally encountered. These and a few less common cultivars are presented in this book. Although Lawson Cypresses are as beautiful as they are variable, their future is uncertain. A deadly introduced root-rot, *Phytophthora lateralis*, is rapidly killing members of this species. The disease is water borne and once present, cannot be eradicated. Infected trees usually die in 1 to 2 years.

Cultivar	Circumference	Height	Crown Spread	AFA Points	Date Last Measured	Location and Nominators
'Alumii'	12'11"	81'	35'	**245**	1989	Woodinville, Chateau Ste Michelle Winery *RVP*
	13'6"	73'	29'	**242**	1989	Woodinville, Chateau Ste Michelle Winery *RVP*
	10'1"	**102'**	29'	230	1995	Olympia, Brewery Mansion *RGB, RVP*
'Aureovariegata'	**7'6"**	**57'**	27'	**154**	1995	Duvall, 16621 W Snoqualmie River Rd *RGB, RVP*
'Columnaris'	**5'2"**	**45'**	11'	**110**	1990	Seattle, Washington Park Arboretum, 19N 5E *RVP*
'Ellwoodii'	–	**46'**	19'	–	1995	Burlington, 735 S Burlington Rd *RVP*
'Erecta Glaucescens'	**7'10"**	**56'**	19'	**155**	1988	Everett, Evergreen Cemetery *RVP*
'Erecta Viridis'	**16'6"**	**83'**	36'	**290**	1987	Sumner, Sumner High School *RVP*
	14'6"	**87'**	32'	269	1987	Puyallup, Valley Ave NW & 122 Ave Ct E *RVP*
'Fletcheri'	–	**39'**	18'	–	1992	Renton, Greenwood Cemetery *RGB,*
	–	38'	17'	–	1988	Bothell, 16323 Simonds Rd *RGB, RVP*
'Fraseri'	**8'7"**	**95'**	25'	**204**	1993	Shelton, 222 Pine St, Simpson Hdqtrs *RVP*
	8'3"	92'	21'	**196**	1992	Puyallup, Woodbine Cemetery *RVP*
'Glauca'	**9'5"**	86'	28'	**206**	1988	Olympia, Brewery Mansion *RVP*
	6'10"	**99'**	37'	190	1988	Tacoma, Wright Park *RVP*
'Gracilis'(?)	**6'1"**	**83'**	31'	**164**	1993	Vashon Island, 16606 99th Ave SW *RGB, RVP*
'Hillieri'	**5'5"**	43'	17'	**112**	1995	Everett, 1725 112th St SW *RGB*
	3'4"	**50'**	16'	94	1995	Bellevue, Highland Community Center *RGB*
'Intertexta'	**3'9"**	**49'**	13'	**98**	1989	Seattle, Green Lake Park, E of pool *ALJ*
'Lutea'	**5'9"**	67'	19'	**141**	1993	Woodland, Hulda Klager Lilac Garden *RVP*
	4'6" 1 of 6 stems	**76'**	20'	135	1993	Snohomish, HD Morgan House, Maple Ave *RGB*

Cultivar	Circumference	Height	Crown Spread	AFA Points	Date Last Measured	Location and Nominators
'Lycopodioides'	**8'9"**	**43'**	31'	**156**	1995	Everett, 515 Laurel Dr *RGB*
'Pendula'	**11'6"**	85'	35'	**232**	1988	Montesano, NW corner of 3rd & Pioneer *RVP*
	7'2"	**104**	21'	195	1992	Puyallup, Woodbine Cemetery *RVP*
'Pottenii'	**5'8"**	**41'**	15'	**113**	1995	Bellevue, 15804 SE 16th St *RGB*
'Stewartii'	**9'0"**	66'	19'	**179**	1992	Renton, Greenwood Cemetery *RGB*
	7'5"	**80'**	33'	177	1993	Bremerton, 4352 Kitsap Lake Rd NW *RGB*
'Tamariscifolia'	**4'6"** 1 of several stems	25'	31'	**87**	1993	Seattle, 2158 E Shelby St *ALJ*
	3'11" 1 of 5 stems	**26'**	33'	81	1993	Tacoma, N 25th near Lawrence *RVP*
'Triomf van Boskoop'	8'3"	**91'**	27'	**197**	1992	Puyallup, Woodbine Cemetery *KVP, RVP*
	8'9"	80'	32'	193	1987	Tacoma, Pt Defiance Park *RVP*
'Versicolor' (?)	**8'2"**	**56'**	24'	**160**	1995	Everett, Evergreen Cemetery *RGB*
'Westermannii'	**9'3"**	**67'**	33'	**186**	1995	Tacoma, Pt Defiance Park *RGB*
'Wisselii'	**6'4"**	**71'**	17'	**151**	1987	Seattle, 47th Ave NE & E Laurelhurst Dr *ALJ*
	6'4"	64'	25'	**146**	1988	Everett, 1516 Rucker *RVP*
'Youngii'(?)	**7'10"** main stem	**67'**	23'	**167**	1993	Longview, 1st Christian Church *RGB, RVP*

Lawson Cypresses come in all shapes and sizes. The largest and one of the most distinct, **Erecta Viridis** is also the oldest. The Puyallup Valley has the best collection of these trees, including our tallest (left). **Wisselii** (right) is also distinct – its dark bluish-green foliage and distinctive silhouette make this one of my favorites. The tree pictured is at a home in Everett.

INTRODUCED TREES

Sawara Cypresses

The Sawara Cypress, *Chamaecyparis pisifera*, from the islands of Japan, is the parent of many common landscape trees and shrubs. Common in old parks and cemeteries, Sawara Cypresses add variety to these large landscapes with a rich array of colors and textures. The variety of leaf form accounts for much of the diversity, and we can find cultivars with scale leaves, needle leaves, thread leaves, and intermediates, most of which are available in green or various shades of yellow.

Plumosa is the largest growing of the Sawaras, and in Washington it has outgrown even the parent species. The tree pictured is our largest Plumosa and is at Sylvester Park in Olympia. Sylvester Park is at the site of the former State Capitol, and is home to several state champion trees. The American Beech on pages 27 and 28 is located in this park and the largest **Box-Elder** is visible in the background of the photo on this page. Sadly, the Box-Elder blew over in a storm in December 1995.

Cultivar	Circumference	Height	Crown Spread	AFA Points	Date Last Measured	Location and Nominators
Chamaecyparis pisifera						
	8'7" largest of 3 stems	67'	40'	**180**	1992	Bellingham, Bayview Cemetery *RGB*
	8'10"	60'	33'	**174**	1992	Bellingham, Bayview Cemetery *RGB*
	5'4"	**73'**	27'	144	1992	Woodinville, Chateau Ste Michelle Winery *RGB, RVP*
'Aurea'	**7'8"**	**66'**	29'	**165**	1992	Seattle, Evergreen Cemetery *RVP*
'Boulevard'	**6'8"**	31'	21'	**116**	1995	Bellevue, Sunset Hills Memorial Park *RGB*
'Filifera'	**9'1"**	63'	23'	**178**	1993	Longview, Huntington Park *RVP*
	8'7"	59'	30'	**169**	1993	Puyallup, 1301 W Stewart Ave *RGB, RVP*
'Filifera Aurea'	**3'0"** largest of 7 trunks	**41'**	27'	**84**	1993	Puyallup, 1301 W Stewart Ave *KVP, RVP*
'Plumosa'	12'5"	85'	33'	**242**	1993	Issaquah, Cascade Savings on Front St *RGB*
	13'1"	75'	36'	**241**	1993	Olympia, Sylvester Park *ALJ, RVP*
	10'0"	**99'**	32'	227	1993	Seattle, Leschi Park *ALJ*
'Plumosa Argentea'	5'4"	**57'**	21'	**126**	1995	Monroe, 421 W Main St *RGB*
	5'6"	44'	24'	116	1995	Seattle, West Seattle Golf Course *RGB*
'Plumosa Aurea'	**10'1"**	67'	27'	**197**	1992	Seattle, Evergreen Park Cemetery *RVP*
	6'6"	**75'**	21'	157	1992	Tacoma, Wright Park near 4th & S G St *RVP*
'Squarrosa'	**12'9"** below forking	49'	36'	**211**	1992	Everett, Cypress Lawn Memorial Park *RGB*
	10'5"	65'	44'	**201**	1987	Olympia, Old Capitol Building *ALJ, RVP*
	5'3"	**73'**	39'	146	1992	Sedro Woolley, N State Multi Service Center *RGB, RVP*

INTRODUCED TREES

	Circumference	Height	Crown Spread	AFA Points	Date Last Measured	Location and Nominators

Pink
Cornus florida f. *rubra*

| 5'9" | 36' | 39' | 115 | 1993 | Washougal, 723 G St *RVP* |

below forking

Silver Pagoda
Cornus alternifolia 'Argentea'

| 2'4" | 24' | 16' | 56 | 1995 | Seattle, 5660 Windermere Rd, tree on NE Keswick *RGB* |

Table
Cornus controversa

| 4'6" | ~35' | 40' | ~99 | 1993 | Whidbey Island, Greenbank, Meerkerk Rhododendron Garden *RGB* |
| 4'1" | 39' | 35' | 97 | 1990 | Seattle, 39th Ave E & E Blaine St *ALJ* |

DOUGLAS-FIR

Bigcone
Pseudotsuga macrocarpa

| 5'7" | 56' | 33' | 131 | 1995 | Seattle, Washington Park Arboretum, 16N 5E *RGB* |

Dwarf
(?) *Pseudotsuga menziesii* 'Slavinii' or affin.

| 7'9" | 58' | 39' | 161 | 1995 | Bellingham, Bayview Cemetery *RGB* |

Weeping
Pseudotsuga menziesii 'Pendula'

| 6'0" | 67' | 31' | 147 | 1995 | Toledo, 376 St Hwy 505 *RGB, RVP* |

DOVE TREE
Davidia involucrata

| 5'4" | 53' | 52' | 130 | 1988 | Snohomish, 1314 Third St *RGB* |
| 4'8" | 55' | 36' | 120 | 1990 | Seattle, 7140 55th Ave S *ALJ, RVP* |

ELDER

European Black
Sambucus nigra

| 4'2" | 28' | 20' | 83 | 1993 | Seattle, 1533 NW 57th St *ALJ* |

ELM

American
Ulmus americana

17'11"	105'	99'	345	1992	Cicero, 14827 300th St NE *RGB* (photo page 50)
13'7"	122'	93'	308	1988	Seattle, 1147 Harvard Ave E *RVP*
13'5"	88'	104'	275	1990	Steilacoom, Western State Hospital *RVP*

Camperdown
Ulmus glabra 'Camperdownii'

| 6'9" | 21' | 35' | 111 | 1993 | Puyallup, 605 7th St SW *RGB, RVP* |

Chinese
Ulmus parvifolia

| 5'4" | 68' | 49' | 144 | 1992 | Redmond, Idylwood Park *RGB* |

Columnar American
Ulmus americana 'Fastigiata'

| 7'10" | 88' | 35' | 191 | 1990 | Tacoma, Calvary Cemetery *KVP, RVP* |

Cornish
Ulmus minor var. *cornubiensis*

| 9'1" | 104' | 43' | 224 | 1995 | Seattle, Rainier Playfield *RGB, RVP* |

English
Ulmus minor var. *vulgaris*

| 14'9" | 114' | 85' | 312 | 1995 | Olympia, Sylvester Park *RVP* |
| 13'0" | 120' | 87' | 298 | 1995 | Seattle, Seattle Pacific University *ALJ* |

European White
Ulmus laevis

15'8"	87'	86'	296	1992	Fall City, 4359 336 Pl SE *RGB* (photo page 50)
10'7"	113'	55'	254	1988	Spokane, Pioneer Park *ALJ, RVP*
11'7"	84'	101'	248	1988	Everett, Evergreen Cemetery *ALJ, RVP*

Guernsey
Ulmus minor var. *sarniensis*

| 13'0" | 102' | 63' | 274 | 1993 | Bellingham, Elizabeth Park *RGB, RVP* |

Huntingdon
(?) *Ulmus* × *hollandica* 'Vegeta'

| 12'8" | 77' | 61' | 244 | 1993 | Walla Walla, 608 W Birch St *Shirley Muse, RGB, RVP* |
| 8'3" | 91' | 51' | 203 | 1990 | Seattle, Harvard Ave E & E Mercer St *ALJ* |

Rock
Ulmus thomasii

| 9'0" | 83' | 69' | 208 | 1993 | Tacoma, Wright Park *ALJ* |
| 7'3" | 90' | 57' | 191 | 1993 | Tacoma, Wright Park *ALJ* |

49

INTRODUCED TREES

Circumference	Height	Crown Spread	AFA Points	Date Last Measured	Location and Nominators
Siberian					*Ulmus pumila*
13'7"	**89'**	**84'**	**273**	1988	Spokane, 4323 3rd Ave E *ALJ, RVP*
14'0"	77'	62'	**260**	1988	Spokane, 4823 3rd Ave E *ALJ, RVP*
Smooth-leaved					*Ulmus minor*
14'2"	97'	**85'**	**288**	1990	Walla Walla, 1107 Alvarado *Shirley Muse, RVP*
13'4"	**103'**	69'	**280**	1990	Walla Walla, 1107 Alvarado *Shirley Muse, RVP*
Tornado					*Ulmus minor* **'Gracilis'**
6'6"	**42'**	**40'**	**130**	1993	Seattle, Washington Park Arboretum, 9N 3W *RVP*
6'10"	29'	26'	118	1993	Spokane, Finch Arboretum *ALJ*
Weeping American					*Ulmus americana* **f. pendula**
10'11"	**75'**	**90'**	**228**	1993	Tacoma, N 6th & N Yakima *KVP, RVP*
Wych					*Ulmus glabra*
14'9"	**118'**	93'	**318**	1995	Tacoma, Wright Park *RVP*
14'10"	104'	**96'**	**306**	1995	Tacoma, Wright Park *RVP*

EMPRESS TREE
Paulownia tomentosa

Circumference	Height	Crown Spread	AFA Points	Date Last Measured	Location and Nominators
12'9"	**48'**	**48'**	**213**	1995	Tacoma, Pt Defiance Park *KVP, RVP*
Lilac					*Paulownia tomentosa* **'Lilacina'**
13'11"	62'	**68'**	**246**	1990	Camas, NE Oak St & NE 5th Ave *RVP*
8'4"	**74'**	55'	188	1995	Seattle, Washington Park Arboretum, 45N 5E *ALJ*

A century ago the **American Elm** was by far the premiere American shade tree. Elm-lined avenues adorned much of New England and the midwest. Today these same streets have no elms. The Dutch Elm disease has eradicated most of this country's American Elm population. The disease has made it to Washington, but is still absent from many areas, as this 100-year-old giant in Cicero can attest.

The European equivalent of our American Elm is the **European White Elm**. While closely related, its appearance is quite distinct. Densely leaved, with sprouting twigs covering much of the branches, it lacks the majestic grace of its American cousin. It is an Elm, however, and is capable of sustained growth leading to impressive dimensions, as the Fall City giant above shows.

INTRODUCED TREES

	Circumference	Height	Crown Spread	AFA Points	Date Last Measured	Location and Nominators

EPAULETTE TREE
Pterostyrax hispida

3'8"	53'	60'	**115**	1993	Seattle, Washington Park Arboretum, 10N 8E *ALJ*

largest of 4 stems

Smooth-barked — *Pterostyrax corymbosa*

4'5"	55'	41'	**118**	1989	Seattle, 1900 Shenandoah Dr E *ALJ*

EUCRYPHIA

Nymans Hybrid — *Eucryphia* x *nymansensis* 'Nymansay'

2'3"	52'	27'	**86**	1995	Bremerton, 1711 Chester *Louise Reh, Billie Barner*

EUODIA

Hupeh — *Tetradium daniellii* (formerly *Euodia hupehensis*)

5'10"	59'	72'	**147**	1988	Seattle, Volunteer Park *ALJ*

Szechwan — *Tetradium daniellii* (formerly *Euodia velutina*)

3'9"	43'	47'	**100**	1995	Seattle, Washington Park Arboretum, 8N 7E *RVP*

FIG

Edible — *Ficus carica*

5'11"	31'	40'	**112**	1987	Seattle, Mt Baker Blvd & 33rd Ave S *ALJ*

FIR

Algerian — *Abies numidica*

4'0"	57'	29'	**112**	1993	Seattle, Washington Park Arboretum, 6N 2E *RVP*

Balsam — *Abies balsamea*

2'7"	67'	20'	**103**	1988	Spokane, Finch Arboretum *ALJ, RVP*
3'1"	42'	19'	84	1992	Tacoma, Calvary Cemetery *KVP, RVP*

Blue Colorado — *Abies concolor* f. *violacea*

8'4"	74'	45'	**185**	1993	Seattle, Acacia Cemetery *RVP*
7'7"	83'	39'	**184**	1993	Sumner, Ryan House, 1228 Main St *RVP*

Blue Spanish — *Abies pinsapo* 'Glauca'

6'7"	71'	23'	**156**	1993	Puyallup, 15014 106th St E *RVP*

Bornmueller — *Abies* x *bornmuelleriana*

8'6"	94'	47'	**208**	1989	Mercer Island, SE 32nd St at 68th Ave SE *RGB*

California Red — *Abies magnifica*

4'5"	79'	17'	**136**	1988	Wind River, USFS Arboretum *RVP*

Caucasian — *Abies nordmanniana*

8'4"	90'	32'	**198**	1990	Longview, Huntington Park *RVP*
9'0"	72'	44'	**191**	1990	Yakima, N 16th Ave & Browne Ave *RVP*
7'0"	92'	33'	184	1990	Allen, 1508 Sam Bell Rd *RVP*

Cilician — *Abies cilicica*

3'0"	54'	21'	**95**	1995	Seattle, Washington Park Arboretum, 16N 5E *RVP*

European Silver — *Abies alba*

5'6"	76'	27'	**149**	1995	Seattle, Washington Park Arboretum, 15N 6E *RVP*
5'8"	68'	30'	**143**	1992	Tacoma, N 18th & N Puget Sound *KVP, RVP*

Fraser — *Abies fraseri*

3'10"	40'	20'	**90**	1995	Bellingham, 3400 Squalicum Pkwy *RGB*
2'9"	46'	17'	83	1995	Bellingham, 3400 Squalicum Pkwy *RGB*

Greek (Apollo) — *Abies cephalonica* var. *graeca*

15'2"	116'	75'	**317**	1993	Tacoma, Pt Defiance Park, entrance *ALJ, RVP* (photo page 52)

Hybrid Caucasian — *Abies* x *insignis*

6'6"	72'	36'	**159**	1993	Seattle, Camp Long, E of pond *ALJ*

INTRODUCED TREES

Circumference	Height	Crown Spread	AFA Points	Date Last Measured	Location and Nominators
Korean					*Abies koreana*
3'2"	32'	23'	**76**	1993	Tukwila, 18010 Southcenter Pkwy *RGB*
2'3"	**43'**	20'	**75**	1990	Wind River, USFS Arboretum *RVP*
Manchurian					*Abies holophylla*
4'1"	**61'**	37'	**119**	1988	Spokane, Finch Arboretum *ALJ, RVP*
Momi					*Abies firma*
7'8"	67'	55'	**173**	1990	Fort Lewis Military Res, N 2nd & Pendleton *RVP*
6'1"	**71'**	32'	152	1990	Tacoma, 4114 N 26th St *RVP*
Moroccan					*Abies pinsapo* ssp. *marocana*
4'6"	**73'**	29'	**134**	1995	Seattle, Washington Park Arboretum, 0 1W *ALJ*
4'10"	68'	27'	**133**	1995	Seattle, Washington Park Arboretum, 0 1W *ALJ*
4'4"	71'	31'	**131**	1995	Seattle, Washington Park Arboretum, 15N 6E *RVP*
Nikko					*Abies homolepis*
6'1"	74'	39'	**157**	1993	Sumner, 0.6 miles N of town on W Valley Hwy *RVP*
6'4"	56'	41'	142	1988	Seattle, 8039 Earl Ave NW *ALJ*
4'3"	**79'**	42'	140	1988	Seattle, Ravenna Park *ALJ*
Pindrow					*Abies pindrow*
4'10"	**64'**	24'	**128**	1995	Seattle, University of Washington, Sieg Hall *ALJ (photo below)*
6'4"	43'	23'	**125**	1988	Seattle, 13th Ave S & S Holgate *ALJ*
Santa Lucia					*Abies bracteata*
3'8"	**57'**	29'	**108**	1990	Seattle, Washington Park Arboretum, 6N 0 *RVP*
4'0"	51'	29'	**106**	1990	Seattle, Washington Park Arboretum, 6N 0 *RVP*

The giant **Greek (Apollo) Fir** in Point Defiance Park in Tacoma is not only the largest introduced fir in the state, but also has the largest branch of any introduced tree. Native to Greece, this and many other Mediterranean trees find our climate favorable.

Not many specimens of the beautiful **Pindrow Fir** can be found in Washington. This is a pity, for it is among our handsomest introduced conifers. Also known as the West Himalayan Fir, this tree makes up part of the mountain forests of India and Nepal.

Circumference	Height	Crown Spread	AFA Points	Date Last Measured	Location and Nominators
Shasta Red					*Abies magnifica* var. *shastensis*
4'2"	**99'**	17'	**153**	1993	Wind River, USFS Arboretum *RVP*
5'1"	54'	23'	121	1988	Seattle, Washington Park Arboretum, 15N 4E *ALJ*
Sicilian					*Abies nebrodensis*
4'0"	**73'**	31'	**129**	1995	Seattle, Washington Park Arboretum, 6N 3E *RVP*
Spanish					*Abies pinsapo*
11'10"	69'	43'	**222**	1995	Bellingham, Carl Cozier Elem School *RGB*
6'10"	**78'**	32'	168	1988	Seattle, 3601 47th Ave NE *ALJ*
Spreading Noble					*Abies procera* 'Glauca Prostrata'
2'5"	**9'**	**23'**	44	1993	Tacoma, Pt Defiance Park *RVP*
Veitch					*Abies veitchii*
2'9"	**84'**	19'	**122**	1993	Seattle, Interlaken Park *ALJ, RVP*
4'7"	57'	24'	**118**	1993	Seattle, 5723 Palatine Ave N *RVP*
White					*Abies concolor*
8'11"	**111'**	38'	**227**	1988	Spokane, NW corner of Havana and Hartson *ALJ, RVP*
9'7"	75'	33'	198	1988	Wenatchee, SW corner of Garfield and First *RVP*

FIRE-TREE

Circumference	Height	Crown Spread	AFA Points	Date Last Measured	Location and Nominators
Chilean					*Embothrium coccineum* var. *lanceolatum*
3'2"	48'	19'	**91**	1989	Seattle, 815 NW 116th St *RGB*
2'9"	**51'**	19'	89	1992	Wauna, 8020 St Hwy 302 *KVP, RGB, RVP*

GINKGO

Circumference	Height	Crown Spread	AFA Points	Date Last Measured	Location and Nominators
					Ginkgo biloba
12'1" below branching	70'	**56'**	229	1990	Walla Walla, 127 Whitman *Shirley Muse*
3'5"	**86'**	21'	132	1988	Tacoma, 601 N Yakima *RVP*

GLORYBOWER

Circumference	Height	Crown Spread	AFA Points	Date Last Measured	Location and Nominators
Harlequin					*Clerodendrum trichotomum*
2'4"	**17'**	**15'**	49	1993	Puyallup, 715 11th St NW *KVP, RVP*

GOLDEN CHAIN

Circumference	Height	Crown Spread	AFA Points	Date Last Measured	Location and Nominators
Alpine					*Laburnum alpinum*
6'9"	**36'**	23'	123	1992	Hoquiam *ALJ*
Chimaeric					+ *Laburnocytisus adamii*
2'8"	**33'**	21'	70	1993	Seattle, Washington Park Arboretum, 15N 5E *ALJ, RVP*
Common					*Laburnum anagyroides*
5'3"	26'	25'	95	1990	Tacoma, Annie Wright Seminary *RVP*
Hybrid					*Laburnum* x *watereri*
5'11"	29'	27'	107	1992	Tacoma, 3705 Olympic Blvd W *RVP*
4'7"	**34'**	27'	96	1990	Seattle, Washington Park Arboretum, 15N 5E *RVP*

GOLDEN RAINTREE

Circumference	Height	Crown Spread	AFA Points	Date Last Measured	Location and Nominators
					Koelreuteria paniculata
5'9"	39'	**47'**	120	1993	Walla Walla, Whitman College *S Muse, RGB, RVP (photo page 54)*
6'0"	33'	30'	112	1990	Tacoma, across S 8th from Grant School *RVP*
5'2"	42'	32'	112	1987	Seattle, Green Lake Park *ALJ*
4'7"	**43'**	45'	109	1993	Walla Walla, Pioneer Park *RGB, RVP*

HACKBERRY

Circumference	Height	Crown Spread	AFA Points	Date Last Measured	Location and Nominators
Common					*Celtis occidentalis*
6'0"	46'	**55'**	132	1988	Walla Walla, Whitman College *RVP*
3'5"	**49'**	33'	98	1988	Spokane, Finch Arboretum *ALJ, RVP*

53

INTRODUCED TREES

	Circumference	Height	Crown Spread	AFA Points	Date Last Measured	Location and Nominators

HARDY RUBBER-TREE
Eucommia ulmoides

	3'11"	53'	33'	108	1990	Seattle, University of Washington, Herb Garden *ALJ*

HAWTHORN (see also **English Hawthorns**, page 56)

Autumn Glory — *Crataegus 'Autumn Glory'*

	2'9"	20'	27'	60	1993	Vancouver, 3512 Main St *RGB, RVP*

Cockspur — *Crataegus crus-galli*

	4'1"	30'	47'	91	1993	Seattle, University of Washington, Medicinal Herb Garden *RVP*
	2'3"	**44'**	23'	77	1988	Seattle, Volunteer Park *ALJ*

Dotted — *Crataegus punctata*

	3'9"	28'	33'	81	1993	Vashon Island, 16820 McIntyre Rd SW *RGB, RVP*
	3'3"	23'	**48'**	74	1993	Seattle, Washington Park Arboretum, 12N 2W *RVP*

Downy — *Crataegus mollis*

	4'8"	38'	44'	105	1993	Spokane, Manito Park, W end of lake *RGB, ALJ*
	3'2"	**45'**	33'	91	1990	Seattle, Volunteer Park *ALJ*

Lavalle — *Crataegus* X *lavallei*

	6'10"	36'	49'	130	1993	Seattle, 3020 44th Ave W *RGB*
	6'6"	38'	41'	126	1993	Seattle, Dunn Garden, 13531 Northshire Rd NW *RGB, RVP*
	5'2"	**43'**	34'	113	1988	Seattle, University of Washington, Anderson Hall *ALJ*

Shining — *Crataegus nitida*

	4'6"	22'	36'	❖85	1988	Seattle, University of Washington, Parrington Hall *ALJ, RVP*

Siberian — *Crataegus sanguinea*

	5'4"	38'	48'	114	1993	Seattle, Volunteer Park *ALJ*

largest of 3 stems

Ron Brightman and Shirley Muse are seen measuring the crown spread of our largest **Golden Raintree**. The tree is located on the Whitman College campus in Walla Walla, home to several other State Champion trees. The Golden Raintree is an attractive small tree that has interesting lantern-like fruits which color nicely in autumn.

INTRODUCED TREES

	Circumference	Height	Crown Spread	AFA Points	Date Last Measured	Location and Nominators
Toba						*Crataegus x mordenensis* 'Toba'
	2'10"	21'	24'	61	1995	Richmond Beach, 1900 N 175th St, Church of the Nazarene *RGB*
Washington						*Crataegus phaenopyrum*
	3'1"	34'	32'	79	1988	Seattle, Volunteer Park *ALJ*
	2'3"	45'	25'	78	1988	Seattle, Volunteer Park *ALJ*
Winter King						*Crataegus viridis* 'Winter King'
	2'8"	29'	33'	69	1995	Seattle, 4325 Densmore Ave N *ALJ*
Yellow Fruited						*Crataegus punctata* 'Aurea'
	3'10"	20'	32'	74	1992	Seattle, 3803 42nd Ave NE *ALJ, RGB*

HAZEL

	Circumference	Height	Crown Spread	AFA Points	Date Last Measured	Location and Nominators
Corkscrew						*Corylus avellana* 'Contorta'
	2'6"	13'	16'	47	1993	Tacoma, Clover Park Arboretum *RVP*
	1'8"	18'	18'	42	1995	Brinnon, Whitney Gardens *RGB, RVP*
European						*Corylus avellana*
	6'6"	34'	49'	124	1993	Vashon Island, Vashon Country Store *Mike Lee, RGB*
Turkish						*Corylus colurna*
	4'6"	58'	39'	122	1988	Seattle, Volunteer Park *ALJ*
	4'11"	43'	45'	113	1992	Vashon Island, SE of Vashon Hwy S & SW 192nd St *RGB*

HEMLOCK

	Circumference	Height	Crown Spread	AFA Points	Date Last Measured	Location and Nominators
Carolina						*Tsuga caroliniana*
	10'6"	74'	59'	215	1990	Tacoma, Pt Defiance Park *RVP* (photo page 58)
Eastern						*Tsuga canadensis*
	11'8"	73'	51'	226	1993	Tacoma, 420 N C St *RVP*
	5'11"	91'	36'	171	1988	Spokane, Pioneer Park *ALJ, RVP*

Iris stands next to our largest **Sargent's Weeping Hemlock**. The tree is located at the Mountain View Cemetery in Tacoma. Not all of our Champion trees are giants, but all are the largest of their type. This hemlock is much larger than any other known out west, although east coast trees over one hundred years old are larger. Sargent's Weeping Hemlocks start life as a ground cover, but after many decades can become small trees.

The **English Hawthorns** are common as street trees in our cities, brightening up otherwise bleak locales. Their ability to withstand harsh city conditions makes them useful in urban landscapes. The trees pictured, outside a Tacoma apartment complex, are runners-up to the trees listed below. **Plena** is the tree on the far left, characterized by double, white flowers which fade to pink. On the right is the ever popular **Paul's Scarlet**, a tree also with double flowers, which are a rich crimson.

English Hawthorns

Many of our ornamental hawthorns are varieties of English Hawthorns. Most cultivars are very old and their exact parentage is complex or unknown. *Crataegus monogyna* (the One-seed or Common Hawthorn) is native to Britain and Europe and is usually distinguished by its single seed. This white-flowered tree is by far the most common type seen in Washington, where it is fully naturalized. This tree and its weeping form are the only English hawthorns with single seeds. The remaining garden forms, which have double and/or colored flowers, may or may not have *monogyna* blood in them and I have put these under *Crataegus laevigata* – the other English, or Midland, Hawthorn, which has 2 or 3 seeds.

Cultivar	Circumference	Height	Crown Spread	AFA Points	Date Last Measured	Location and Nominators
Crataegus monogyna						
	9'3"	37'	58'	❖163	1992	Mount Vernon, 1861 Dike Rd *RGB*
	5'7"	67'	37'	143	1995	Seattle, Volunteer Park *ALJ*
f. *pendula*	7'0"	38'	36'	131	1987	Bellingham, 2527 Broadway *RVP*
Crataegus laevigata						
'Bicolor'	5'2"	30'	40'	102	1993	Tacoma, 4202 N 16th *KVP, RVP*
	4'11"	31'	45'	101	1993	Seattle, 607 18th Ave E *RVP*
'Crimson Cloud'	2'1"	23'	17'	52	1995	Yakima, Yakima Area Arboretum *RGB*
'Masekii'	7'8"	42'	47'	146	1993	Toppenish, Park at N D St & Lincoln Ave *RGB, RVP*
'Paul's Scarlet'	7'1"	37'	43'	133	1993	Walla Walla, 835 Whitman St *RGB, RVP*
'Plena'	4'2"	34'	29'	91	1993	Seattle, 403 16th Ave E *ALJ*
	4'0"	30'	31'	86	1993	Tacoma, 515 N 2nd St *KVP, RVP*
'Punicea'	4'3"	32'	43'	94	1993	Seattle, Green Lake Park, near Corliss St *ALJ*
'Rosea Flore Pleno'	4'10"	37'	31'	103	1993	Tacoma, 2202 N Ferdinand *KVP, RVP*

Circumference	Height	Crown Spread	AFA Points	Date Last Measured	Location and Nominators
Jenkins'					*Tsuga canadensis* 'Jenkinsii'
2'4"	**48'**	**25'**	**82**	1993	Seattle, Washington Park Arboretum, 27N 3E *ALJ*
Northern Japanese					*Tsuga diversifolia*
3'3"	**34'**	**30'**	**80**	1992	Richmond Beach, 714 NW 189th Ln *RGB*
Sargent's Weeping					*Tsuga canadensis* 'Pendula'
4'6"	**9'**	**24'**	**69**	1993	Tacoma, Mountain View Cemetery *KVP, RVP (photo page 55)*
Southern Japanese					*Tsuga sieboldii*
1'10"	**48'**	**27'**	**77**	1993	Seattle, Washington Park Arboretum, 30N 2E *ALJ*
best of 2 stems					

HICKORY

Circumference	Height	Crown Spread	AFA Points	Date Last Measured	Location and Nominators
Mockernut					*Carya tomentosa*
7'5"	**89'**	**72'**	**196**	1987	Tacoma, Wright Park *RVP (photo page 58)*
Pignut					*Carya glabra*
6'9"	**92'**	**61'**	**188**	1987	Steilacoom, Western State Hospital *RVP*
Red					*Carya ovalis*
10'4"	**96'**	**66'**	**236**	1992	Prairie, 1996 Prairie Rd, Warner's Prairie *RGB*
Shagbark					*Carya ovata*
8'4"	87'	**51'**	**200**	1992	Prairie, 1996 Prairie Rd, Warner's Prairie *RGB*
7'7"	**92'**	**55'**	**197**	1992	Prairie, 1996 Prairie Rd, Warner's Prairie *RGB*
7'4"	90'	**54'**	**191**	1992	Prairie, 1996 Prairie Rd, Warner's Prairie *RGB*
Shellbark					*Carya laciniosa*
6'0"	**77'**	**52'**	**162**	1993	Sultan, 106 4th Ave *RGB, RVP*

HOLLY

Circumference	Height	Crown Spread	AFA Points	Date Last Measured	Location and Nominators
American					*Ilex opaca*
3'6"	28'	**27'**	**77**	1995	Bellingham, 1400 Yew St *RGB, RVP*
2'5"	**41'**	20'	**75**	1992	Seattle, 4221 E Lee St *ALJ, RVP*
Balearic					*Ilex* x *altaclerensis* 'Balearica'
5'6"	**27'**	**27'**	**100**	1995	Anacortes, Causland Park *RGB*
Camellia-leaved Highclere					*Ilex* x *altaclerensis* 'Camelliaefolia'
4'0"	55'	19'	**108**	1990	Seattle, 908 14th Ave E *ALJ*
4'4"	50'	19'	**107**	1990	Seattle, 908 14th Ave E *ALJ*
2'9"	**62'**	14'	98	1992	Lakewood, Lakewold Gardens *KVP, RVP*
best of 2 stems					
English					*Ilex aquifolium*
12'0"	50'	45'	**205**	1993	Elma, 201 E Young St *RGB*
fused base					
4'4"	**58'**	**48'**	122	1987	Seattle, Volunteer Park *ALJ*
largest of 29 stems					
Golden King Highclere					*Ilex* x *altaclerensis* 'Golden King'
4'2"	**30'**	19'	**85**	1992	Vashon Island, 22024 Monument Rd SW *Mike Lee*
Golden Queen					*Ilex aquifolium* 'Golden Queen'
2'1"	**19'**	**16'**	**48**	1995	Anacortes, Causland Park *RGB*
Gold Variegated					(?) *Ilex aquifolium* 'Aurea Marginata'
4'10"	**26'**	**21'**	**89**	1991	Ferndale, 2069 Vista *RGB*
Hedgehog					*Ilex aquifolium* 'Ferox'
2'10"	**27'**	**20'**	**65**	1993	Longview, Huntington Park *RVP*
Hodgins					*Ilex* x *altaclerensis* 'Hodginsii'
3'5"	**31'**	**15'**	**76**	1993	Tacoma, Old Tacoma Cemetery *RVP*
Japanese					*Ilex crenata*
1'6"	18'	**19'**	**41**	1992	Lakewood, Lakewold Gardens *RGB, RVP*
1'0"	**24'**	9'	**38**	1993	Seattle, Washington Park Arboretum, 6N 3E *RVP*

INTRODUCED TREES

Circumference	Height	Crown Spread	AFA Points	Date Last Measured	Location and Nominators
Orange-berried					*Ilex aquifolium* **'Amber'**
3'5"	**38'**	20'	**84**	1995	Lynnwood, 164th & 22nd Ave W *RGB*
2'9"	33'	**23'**	72	1995	Bellingham, 336 N Forest St *RGB*
Perry's Silver Weeping					*Ilex aquifolium* **'Argentea Marginata Pendula'**
3'6"	**13'**	16'	**59**	1995	Tacoma, 1615 6th Ave *RGB*
2'11"	**13'**	**18'**	52	1992	Vashon Island, Vashon Cemetery *Mike Lee, RGB, RVP*
Red Twigged					*Ilex aquifolium* **'Rubricaulis Aurea'**
2'9"	**25'**	**19'**	**63**	1995	Mt Vernon, 917 S 3rd *RGB*
Silver Hedgehog					*Ilex aquifolium* **'Ferox Argentea'**
3'0"	**26'**	**19'**	**67**	1995	Mt Vernon, 917 S 3rd *RGB*
Silver Variegated					*Ilex aquifolium* **'Argentea Marginata'**
4'8"	39'	**25'**	**101**	1995	Aberdeen, Sam Benn Park *Neil Leonard, RGB*
2'1"	**48'**	17'	77	1995	Bellingham, Cornwall Park, SE entrance *RGB*
Yellow-berried					*Ilex aquifolium* **f. bacciflava**
3'8"	**38'**	**20'**	**87**	1990	Tacoma, 715 N J St *RVP*

HONEYLOCUST (see also LOCUST, page 64)   *Gleditsia triacanthos*

Circumference	Height	Crown Spread	AFA Points	Date Last Measured	Location and Nominators
7'6"	**88'**	**57'**	**192**	1993	Washougal, 713 G St *RVP*
Globe					*Gleditsia triacanthos* **'Elegantissima'**
3'0"	**38'**	**37'**	**83**	1995	Seattle, University of Washington, Benson Hall *RGB*
Ruby Lace					*Gleditsia triacanthos* **'Ruby Lace'**
1'10"	**31'**	**23'**	**59**	1993	Puyallup, 10825 57th St E *KVP, RVP*
Sunburst					*Gleditsia triacanthos* **'Sunburst'**
5'8"	**54'**	**63'**	**138**	1993	Walla Walla, 625 Sumach St *RGB, RVP*
5'10"	49'	47'	131	1993	Dayton, 309 S 3rd St *Shirley Muse, RGB, RVP*

A very rare hemlock related to our Mountain Hemlock is the **Carolina Hemlock**, found only in a few locations in the southern Appalachian Mountains. This attractive tree in Point Defiance Park in Tacoma is nearly as large as those found in the wild.

Hickories are uncommon in Washington, which is a pity, since they are glorious trees. Their bold leaves turning bright yellow in autumn are a joy to behold. Shown above is the state's largest **Mockernut Hickory**, which is in Wright Park in Tacoma.

Circumference	Height	Crown Spread	AFA Points	Date Last Measured	Location and Nominators
Thornless					*Gleditsia triacanthos* **f. inermis**
7'10"	66'	**67'**	**177**	1995	Richland, Howard Amon Park *Mid-Columbia Forestry Council*
8'3"	56'	**67'**	**172**	1993	Walla Walla, 911 4th Ave S, on Willard St *RGB, RVP*
7'1"	71'	63'	**172**	1987	Seattle, Viretta Park *ALJ (photo below)*
5'6"	**90'**	49'	168	1987	Redmond, Marymoor Park *ALJ, RVP*

HOP-HORNBEAM

Eastern					*Ostrya virginiana*
2'4"	62'	29'	97	1988	Seattle, Lincoln Park *ALJ*
European					*Ostrya carpinifolia*
4'4"	66'	**51'**	131	1990	Seattle, Washington Park Arboretum, 46N 7E *RVP*

HORNBEAM

American					*Carpinus caroliniana*
3'3"	**23'**	**54'**	76	1993	Seattle, Washington Park Arboretum, 45N 7E *ALJ, RVP*
3'6"	14'	35'	65	1993	Seattle, Washington Park Arboretum, 46N 7E *ALJ, RVP*
European					*Carpinus betulus*
10'8"	74'	66'	**218**	1992	Auburn, 506 1st St SW *RGB (photo below)*
7'7"	73'	**76'**	183	1990	Seattle, 310 39th Ave E *ALJ*
Looseflower					*Carpinus laxiflora*
3'9"	26'	**33'**	79	1993	Seattle, University of Washington, Raitt Hall *ALJ, RVP*

Close relatives of the acacias of the African Savanna, and just as armored, are the honeylocusts. However, the commonly planted types are without thorns, like Seattle's champion **Thornless Honeylocust**, pictured here.

Hornbeams are generally small-growing trees with very hard wood that is commonly used for tool handles and other small implements. The **European Hornbeam**, however, can get quite large, as this Auburn tree demonstrates.

INTRODUCED TREES

	Circumference	Height	Crown Spread	AFA Points	Date Last Measured	Location and Nominators
Oriental						*Carpinus orientalis*
	3'9"	**35'**	**45'**	**91**	1993	Seattle, Washington Park Arboretum, 45N 7E *ALJ*
Pyramidal						*Carpinus betulus* **'Fastigiata'**
	3'4"	**57'**	26'	**103**	1988	Seattle, 2701 3rd Ave W *RVP*
	3'8"	36'	31'	88	1987	Seattle, Washington Park Arboretum, 47N 8E *ALJ*

HORSECHESTNUT (see also BUCKEYE, page 32)

	Circumference	Height	Crown Spread	AFA Points	Date Last Measured	Location and Nominators
Briot						*Aesculus* x *carnea* **'Briotii'**
	4'11"	29'	**41'**	**98**	1992	Tacoma, 1705 Sunset Dr W *KVP, RVP*
	3'10"	**38'**	33'	92	1993	Tacoma, 8201 S 17th St *KVP, RVP*
Common						*Aesculus hippocastanum*
	14'6"	83'	79'	**277**	1988	Walla Walla, Pioneer Park *RVP*
	14'7"	73'	60'	262	1987	Seattle, 922 28th Ave S *ALJ (photo below)*
	9'10"	**98'**	62'	231	1995	Tacoma, Wright Park *RVP*
	12'3"	56'	**89'**	223	1990	Tacoma, Pt Defiance Park *RVP*
Double Flowered						*Aesculus hippocastanum* **'Baumannii'**
	15'2"	79'	62'	**277**	1987	Seattle, 722 28th Ave S *ALJ*
	11'5"	**83'**	67'	232	1988	Spokane, Pioneer Park *ALJ, RVP*
Indian						*Aesculus indica*
	5'7"	**35'**	**30'**	**109**	1992	Seattle, Carl S English Gardens, Ballard Locks *ALJ, RVP*
Japanese						*Aesculus turbinata*
	6'4"	**37'**	**34'**	**121**	1990	Seattle, E Jefferson St W of 22nd Ave *ALJ*

One of few large-growing trees with spectacular floral displays is the **Common Horsechestnut**. Native to Europe, this robust, shade-producing tree makes the non-edible chestnuts, or 'conkers' that many of us are familiar with. This giant in Seattle is among our largest.

Much planted as a small shrub in the Pacific Northwest for its unusual growth habit, the **Hollywood Juniper** can grow into a tree. Washington's largest, pictured above, is 27 feet tall, and others taller still grow in California.

Circumference	Height	Crown Spread	AFA Points	Date Last Measured	Location and Nominators
Red					*Aesculus* × *carnea*
9'8"	48'	51'	**177**	1993	Tacoma, 707 E 65th *KVP, RVP*
7'11"	62'	**67'**	**174**	1988	Seattle, Me-Kwa-Mooks Park *ALJ*
5'6"	**69'**	43'	146	1988	Seattle, Volunteer Park *ALJ*
Sunrise					*Aesculus* × *neglecta* 'Erythroblastos'
2'2"	**39'**	21'	70	1991	Seattle, 1900 Shenandoah Dr E *RGB*

IRONWOOD

Circumference	Height	Crown Spread	AFA Points	Date Last Measured	Location and Nominators
Persian					*Parrotia persica*
3'2"	**60'**	**34'**	**106**	1990	Lakewood, Lakewold Gardens *KVP, RVP*

JUNIPER

Circumference	Height	Crown Spread	AFA Points	Date Last Measured	Location and Nominators
Blue Eastern					*Juniperus virginiana* **f. glauca**
5'7"	**44'**	25'	**117**	1993	Tacoma, Mountain View Cemetery *RVP*
5'1"	39'	**31'**	108	1993	Seattle, NE 52nd St at Brooklyn Ave NE *ALJ*
Chinese					*Juniperus chinensis*
4'6"	51'	25'	**113**	1988	Wenatchee, 206 S Miller *RVP*
4'9"	46'	23'	**111**	1988	Seattle, 7051 Ravenna Ave *ALJ*
3'3"	**60'**	18'	103	1988	Spokane, Finch Arboretum *ALJ, RVP*
Farges'					*Juniperus squamata* **var. fargesii**
5'0"	38'	**27'**	105	1993	Seattle, Golden Gardens Park *RVP*
Hollywood					*Juniperus chinensis* **var. torulosa**
4'0"	27'	21'	**80**	1993	Tacoma, N 28th & Junett *RVP* (photo page 60)
2'5"	**31'**	20'	65	1993	Seattle, S Holly and 57th Ave S *ALJ*
3'0"	25'	**24'**	67	1992	Seattle, 4115 Brooklyn Ave NE *ALJ*
Irish					*Juniperus communis* 'Hibernica'
2'2"	**30'**	12'	59	1995	Bryant, 26832 35th Ave NE *RGB*
Keteleer					*Juniperus chinensis* 'Keteleerii'
4'8"	**40'**	25'	**102**	1988	Seattle, Washington Park Arboretum, 52N 7W *ALJ*
Meyer					*Juniperus squamata* 'Meyeri'
5'6"	29'	**39'**	105	1993	Tacoma, New Tacoma Cemetery *RVP*
below forking					*Juniperus chinensis* 'Pyramidalis'
Pyramidal Chinese					
3'3"	**32'**	17'	75	1995	Arlington, 210 Hamlin Dr *RGB*
3'3"	27'	20'	71	1995	Tacoma, 511 N K St *RVP*
Variegated Hollywood					*Juniperus chinensis* **var. torulosa** 'Variegata'
4'7"	28'	**24'**	89	1993	Burlington, Cemetery at Gardner Rd & Allston Ln *RGB*
Weeping Eastern					*Juniperus virginiana* 'Pendula'
4'9"	42'	37'	108	1993	Dayton, 518 S 1st St *Shirley Muse, RGB, RVP*
Young's Golden					*Juniperus chinensis* 'Aurea'
1'1"	**27'**	12'	43	1995	Seattle, University of Washington, Medicinal Herb Garden *RGB*

KATSURA

Circumference	Height	Crown Spread	AFA Points	Date Last Measured	Location and Nominators
					Cercidiphyllum japonicum
10'11"	84'	**63'**	**231**	1993	Seattle, 1900 Shenandoah Dr E *ALJ*
5'7"	**97'**	52'	177	1995	Seattle, Washington Park Arboretum, 13N 8E *ALJ*
Magnificent					*Cercidiphyllum magnificum*
4'8"	**43'**	**41'**	**109**	1992	Seattle, Washington Park Arboretum, 30N 3E *ALJ*

KENTUCKY COFFEE TREE

Circumference	Height	Crown Spread	AFA Points	Date Last Measured	Location and Nominators
					Gymnocladus dioicus
5'0"	**62'**	48'	**134**	1993	Walla Walla, 1124 Alvarado St *RGB, RVP*
4'1"	59'	**56'**	122	1993	Seattle, Volunteer Park *ALJ*

Circumference	Height	Crown Spread	AFA Points	Date Last Measured	Location and Nominators
LARCH					
Dunkeld					*Larix* x *eurolepis*
9'3"	**69'**	52'	**193**	1993	Tacoma, Pt Defiance Park *RVP*
European					*Larix decidua*
10'10"	95'	55'	**239**	1988	Spokane, Coeur d'Alene Park *ALJ, RVP (photo below)*
13'0"	46'	54'	215	1993	Ferndale, cemetery at W Axton Rd & Northwest Dr *RGB*
7'1"	**109'**	41'	204	1988	Spokane, Coeur d'Alene Park *ALJ, RVP*
Golden					*Pseudolarix amabilis*
2'4"	**25'**	21'	**58**	1995	Seattle, Kubota Gardens Park *RGB*
Japanese					*Larix kaempferi*
4'6"	**88'**	33'	**150**	1993	Seattle, Ravenna Park *ALJ*
5'3"	35'	47'	110	1995	Anacortes, Causland Park *RGB, RVP*
Polish					*Larix decidua* ssp. *polonica*
3'9"	**79'**	23'	**130**	1988	Wind River, USFS Arboretum *RVP*
4'7"	64'	33'	**127**	1988	Seattle, Washington Park Arboretum, 19N 2W *RVP*
Tamarack					*Larix laricina*
4'5"	**56'**	35'	**118**	1988	Seattle, Washington Park Arboretum, 18N 2W *RVP*
Weeping European					*Larix decidua* 'Pendula'
3'2"	**23'**	19'	**66**	1995	Seattle, Volunteer Park, along W fence *RGB*
LAUREL					
Bay					*Laurus nobilis*
1'9"	**41'**	23'	**68**	1993	Seattle, University of Washington, Sieg Hall *ALJ*
largest of 5 stems					

The **European Larch** is one of our more commonly planted species of this group of deciduous conifers. This tree, found in Spokane's Coeur d' Alene Park, is our largest introduced Larch.

Laurels are commonly planted as shrubs. If left unattended, they can grow to tree size. This **Portugal Laurel** at the Foster Golf Links in Tukwila, largest in the state, proves the point.

Circumference	Height	Crown Spread	AFA Points	Date Last Measured	Location and Nominators
English					*Prunus laurocerasus*
8'0"	32'	43'	❖**139**	1988	Seattle Center Flag Pavillion *ALJ*
2'6"	**44'**	**48'**	86	1990	Seattle, Golden Gardens Park *RVP*
Magnolia Leaved					*Prunus laurocerasus* **'Magnoliaefolia'**
3'10"	**38'**	**64'**	**100**	1993	Seattle, Washington Park Arboretum, 46N 2E *RGB*
Portugal					*Prunus lusitanica*
12'6"	45'	**51'**	**208**	1992	Tukwila, 13500 Interurban, Foster Golf Links *RGB (photo page 62)*
3'4"	**59'**	35'	108	1992	Lakewood, Lakewold Gardens *RGB, RVP*
largest of 2 stems					

LILAC

Circumference	Height	Crown Spread	AFA Points	Date Last Measured	Location and Nominators
Japanese Tree					*Syringa reticulata*
5'1"	**25'**	**29'**	**93**	1992	Sultan, 705 3rd St *RGB, ALJ*

LINDEN (see also **BASSWOOD**, page 28)

Circumference	Height	Crown Spread	AFA Points	Date Last Measured	Location and Nominators
Bigleaf					*Tilia platyphyllos*
12'3"	83'	60'	**245**	1992	Acme, 1424 St Rt 9 *RGB*
7'2"	**96'**	57'	196	1987	Tacoma, Wright Park *RVP*
8'0"	83'	**67'**	203	1987	Seattle, Kinnear Park *ALJ*
Crimean					*Tilia* **'Euchlora'**
7'6"	59'	**44'**	**160**	1988	Seattle, W Raye St across from cemetery *ALJ*
4'9"	**69'**	34'	134	1988	Seattle, street tree at Mt Pleasant Cemetery *ALJ*
European					*Tilia* x *europaea*
10'8"	**80'**	**64'**	**224**	1993	Toppenish, Park at N D St & Lincoln Ave *RGB, RVP*

The **Littleleaf Linden**, a European cousin to our Basswoods, is our largest and most planted member of the genus *Tilia*. The biggest of all is this specimen growing in the rich soils of the Skagit River Valley in Avon.

INTRODUCED TREES

Circumference	Height	Crown Spread	AFA Points	Date Last Measured	Location and Nominators
Littleleaf					*Tilia cordata*
14'9"	81'	**79'**	**278**	1988	Avon, Barrett Rd & Avon-Allen Rd *RVP* (photo page 63)
14'0"	88'	60'	**271**	1989	Seattle, Denny-Blaine Park *ALJ*
13'0"	**96'**	60'	267	1992	Arlington, Burn Rd S of town *RGB, RVP*
Silver					*Tilia tomentosa*
13'9" below forking	77'	**83'**	**263**	1993	Seattle, 21st Ave NE & NE 50th St *ALJ*
11'1"	**78'**	67'	229	1993	College Place, Walla Walla College *Shirley Muse, RGB, RVP*
Silver Pendent					*Tilia tomentosa* 'Pendula'
9'2"	79'	**58'**	**203**	1993	Tacoma, 5002 Asotin *RVP*
7'7"	**84'**	52'	188	1990	Woodinville, Chateau Ste Michelle Winery *RVP*
Spectacular					*Tilia* x *spectabilis*
7'7"	**69'**	**60'**	**179**	1990	Seattle, Kinnear Park *ALJ*

LOCUST (see also HONEYLOCUST, page 58)

Circumference	Height	Crown Spread	AFA Points	Date Last Measured	Location and Nominators
Black					*Robinia pseudoacacia*
20'3"	95'	**76'**	**357**	1987	Auburn, 17208 Green Valley Rd *RVP*
14'9"	**109'**	74'	304	1993	Walla Walla, Whitman College *Shirley Muse* (photo below)
Columnar					*Robinia pseudoacacia* 'Pyramidalis'
4'8"	**67'**	25'	**129**	1992	Seattle, Washington Park Arboretum, 16N 6E *RVP*
Contorted					*Robinia pseudoacacia* 'Tortuosa'
4'8"	**42'**	**31'**	**106**	1993	Puyallup, 1102 15th St NW *KVP, RVP*
Globe					*Robinia pseudoacacia* 'Umbraculifera'
7'4"	**38'**	23'	**132**	1993	Walla Walla, Whitman College *Shirley Muse, RGB, RVP*
5'8"	26'	**32'**	102	1990	Tacoma, 6035 S Pine St *RVP*

The towering **Black Locust** pictured here is at the Whitman College campus in Walla Walla. Black locusts were commonly planted by pioneers for their stong, fast-growing wood. Today, Black Locusts may be found in nearly every state.

The small, fruit-producing **Loquat**, common in warmer climes, is rare in Washington. To see one survive long enough to become a tree, as this specimen in Seattle has, is astounding.

Circumference	Height	Crown Spread	AFA Points	Date Last Measured	Location and Nominators
Golden					*Robinia pseudoacacia* 'Frisia'
7'6"	**72'**	44'	**173**	1992	Auburn, 625 M St *RGB*
Hybrid Mexican					*Robinia* x *holdtii*
7'4"	**61'**	57'	**163**	1995	Seattle, Washington Park Arboretum, 17N 7E *RVP*
Idaho					*Robinia* x *ambigua* 'Idahoensis'
4'7"	39'	**41'**	**104**	1993	Seattle, Washington Park Arboretum, 15N 5E *RVP*
4'0"	**49'**	25'	103	1992	Steilacoom, Western Washington State Hospital *KVP, RVP*
Pink					*Robinia* x *ambigua* 'Decaisneana'
8'11"	68'	47'	**187**	1988	Seattle, Monterey Pl S, E of Beacon Ave S *RVP*
6'5"	**73'**	**51'**	163	1995	Seattle, Washington Park Arboretum, 17N 7E *RVP*
Singleleaf					*Robinia pseudoacacia* 'Unifoliola'
5'1"	**65'**	**37'**	**135**	1995	Seattle, Washington Park Arboretum, 17N 7E *RVP*
LOQUAT					*Eriobotrya japonica*
3'4"	**29'**	**29'**	76	1989	Seattle, 612 NW 56th St *ALJ* (photo page 64)
MAACKIA					
Chinese					*Maackia chinensis*
5'5"	**52'**	44'	**128**	1995	Seattle, Lake Washington Blvd E & E Howell St *RGB, ALJ*
5'8"	40'	**49'**	120	1994	Seattle, Washington Park Arboretum, 17N 5E *RVP*
Japanese					*Maackia amurensis* var. *buergeri*
7'10"	**49'**	**56'**	**157**	1989	Seattle, Washington Park Arboretum, 17N 5E *RVP*
MAGNOLIA					
Alexandrina					*Magnolia* x *soulangiana* 'Alexandrina'
4'5" below branching	25'	29'	85	1993	Tacoma, 4010 N 34th *KVP, RVP*
3'8"	**27'**	24'	77	1993	Tacoma, 7030 A St *KVP, RVP*
3'2"	26'	**33'**	72	1993	Tacoma, 4133 S 7th *KVP, RVP*
Anhwei					*Magnolia cylindrica*
1'8"	**33'**	27'	59	1995	Seattle, Washington Park Arboretum, 26N 0 *RGB*
Bigleaf					*Magnolia macrophylla*
5'3"	**54'**	40'	**127**	1993	Seattle, 625 34th Ave E *RGB*
3'8"	52'	**45'**	107	1990	Seattle, Washington Park Arboretum, 28N 1E *ALJ*
Brozzoni					*Magnolia* x *soulangiana* 'Brozzonii'
3'6"	35'	**38'**	87	1993	Seattle, Washington Park Arboretum, *RVP*
Campbell					*Magnolia campbellii*
5'4"	47'	**41'**	**121**	1993	Seattle, Washington Park Arboretum, 11N 7E *ALJ*
5'11"	37'	35'	**117**	1993	Seattle, 615 W Highland Dr *ALJ*
3'1"	**64'**	31'	109	1988	Seattle, Washington Park Arboretum, 29N 2E *ALJ*
Cucumber Tree					*Magnolia acuminata*
7'10"	**82'**	**71'**	**194**	1993	Seattle, 13531 Northshire Rd NW, Dunn Garden *RGB, ALJ*
Dawson					*Magnolia dawsoniana*
4'5"	61'	40'	**124**	1993	Seattle, Washington Park Arboretum, 28N 2E *RVP*
4'5"	58'	**45'**	122	1990	Seattle, 1900 Shenandoah Dr E *RGB*
3'1"	**75'**	27'	119	1994	Seattle, Washington Park Arboretum, 28N 2E *ALJ*
Evergreen					*Magnolia grandiflora*
6'8"	35'	32'	**123**	1990	Woodland, Hulda Klager Lilac Gardens *RVP*
5'10"	**44'**	32'	122	1995	Seattle, 5011 9th Ave E *Mike Lee*
Fraser					*Magnolia fraseri*
4'0"	**78'**	**40'**	**136**	1993	Seattle, Washington Park Arboretum, 23N 1E *ALJ*

INTRODUCED TREES

Circumference	Height	Crown Spread	AFA Points	Date Last Measured	Location and Nominators
Goliath					*Magnolia grandiflora* **'Goliath'**
4'4"	**26'**	**24'**	**84**	1992	Seattle, 1900 Shenandoah Dr E *RGB*
Japanese Silverleaf					*Magnolia hypoleuca*
8'5" below forking	46'	43'	**158**	1992	Tacoma, University of Puget Sound, S lawn area *RVP*
4'7"	**64'**	**52'**	132	1988	Tacoma, Wright Park *ALJ, RVP*
Kobus					*Magnolia kobus*
7'7"	49'	55'	**154**	1989	Edmonds, 711 4th Ave S *RGB*
7'4"	52'	52'	**153**	1992	Puyallup, 7321 Stewart Ave E *RGB, RVP*
4'3"	**60'**	52'	124	1990	Seattle, Washington Park Arboretum, 22N 0 *ALJ*
5'7"	58'	**59'**	140	1992	Avondale, Quail Hollow Farm, 12653 Avondale Rd *RGB*
Merrill					*Magnolia kobus* var. *loebneri* **'Merrill'**
3'10"	**29'**	**24'**	**81**	1995	Seattle, 3806 E McGilvra St *RGB*
Pink Star					*Magnolia stellata* **'Rosea'**
5'1"	22'	**28'**	**90**	1993	Puyallup, 7300 Stewart Ave E *RGB*
2'5"	**28'**	25'	63	1992	Seattle, 1900 Shenandoah Dr E *RGB*
Pyramid					*Magnolia fraseri* ssp. *pyramidata*
3'3"	**52'**	**29'**	**98**	1993	Dayton, 519 S 1st St *RGB*
Rustica Rubra					*Magnolia* x *soulangiana* **'Rustica Rubra'**
4'11"	28'	**33'**	**95**	1993	Seattle, 3915 48th Pl NE *ALJ*
2'5"	**31'**	24'	66	1993	Tacoma, SW corner N C St & N 7th St *RVP*
Sargent					*Magnolia sargentiana* var. *robusta*
7'1"	48'	44'	**144**	1990	Seattle, 1900 Shenandoah Dr E *ALJ*
4'0"	**71'**	49'	131	1994	Seattle, Washington Park Arboretum, 29N 2E *ALJ*
5'3"	59'	**53'**	135	1993	Seattle, 13531 Northshire Rd NW, Dunn Garden *RGB, RVP*

The **Veitch Magnolia** is one of the many large, deciduous magnolias that are spectacular when in bloom. First bred in 1907, it is a cross between the Campbell and Yulan Magnolias. I had seen only smallish trees of this type when Arthur Jacobson discovered this tree growing in northeast Seattle. The rugged form, big leaves, and large, pink flowers make this tree worth a visit during its spring bloom. Although not as big as the largest in England, the tree pictured may be the largest in North America.

Circumference	Height	Crown Spread	AFA Points	Date Last Measured	Location and Nominators
Saucer					*Magnolia* x *soulangiana*
5'7"	25'	**39'**	**102**	1988	Seattle, 24th Ave NW, N of NW 70th St *ALJ*
3'1"	**39'**	38'	85	1989	Seattle, Washington Park Arboretum, 28N 3E *ALJ, RVP*
Sprenger					*Magnolia sprengeri*
3'0"	**53'**	33'	97	1992	Seattle, Washington Park Arboretum, 11N 7E *RGB*
Sweet Bay					*Magnolia virginiana*
4'2"	60'	25'	**118**	1988	Seattle, 203 Lake Washington Blvd E *Mike Lee*
Thompson					*Magnolia* x *thompsoniana*
1'11"	24'	20'	50	1992	Richmond Beach, 714 NW 189th Ln *RGB*
Umbrella					*Magnolia tripetala*
3'7"	58'	**45'**	112	1989	Seattle, Washington Park Arboretum, 11N 7E *ALJ*
4'11"	29'	35'	97	1993	Seattle, 1707 NW 63rd St *KVP, RVP*
2'8"	**64'**	27'	103	1989	Seattle, Washington Park Arboretum, 28N 2E *ALJ*
Veitch					*Magnolia* x *veitchii*
6'2"	48'	**51'**	**135**	1990	Seattle, 10019 48th Ave NE *RGB* (*photo page 66*)
3'5"	**57'**	33'	109	1995	Seattle, Washington Park Arboretum, 14N 6E *RVP*
Victoria					*Magnolia grandiflora* 'Victoria'
6'5"	38'	32'	**123**	1995	Seattle, 5660 NE Windermere, tree on Keswick NE *RGB*
4'4"	39'	35'	100	1995	Puyallup, 2027 7th Ave SE *KVP, RVP*
Wada's Memory					*Magnolia* x *proctoriana* 'Wada's Memory'
4'4"	42'	31'	**102**	1992	Seattle, Washington Park Arboretum, 11N 7E *ALJ*
3'0" below	**47'**	**35'**	92	1993	Seattle, Washington Park Arboretum, 11N 8E *RVP*
largest stem					
Watson					*Magnolia* x *wieseneri*
3'6"	**27'**	23'	**75**	1995	Seattle, 1728 45th Ave SW *RGB*
3'7"	24'	**29'**	74	1993	Seattle, 9750 45th Ave SW, Johnson Rhododendron Garden *ALJ*
2'5"	**27'**	23'	62	1993	Centralia, 721 W First St *RGB, RVP*
White Saucer					*Magnolia* x *soulangiana* f. *alba*
4'10"	**23'**	25'	87	1993	Tacoma, 1921 Martin Luther King Dr *KVP, RVP*
below branching					
Willowleaf					*Magnolia salicifolia*
3'1"	**55'**	41'	**102**	1993	Seattle, Washington Park Arboretum, 22N 2E *ALJ*
Yellow Cucumber Tree					*Magnolia acuminata* var. *subcordata*
3'1"	**38'**	25'	81	1992	Seattle, 1002 37th Ave E *ALJ*
Yulan					*Magnolia denudata*
8'6"	40'	39'	**152**	1990	Woodland, Hulda Klager Lilac Gardens *RGB*
below forking					
3'11"	**55'**	44'	113	1990	Seattle, Washington Park Arboretum, 28N 2E *ALJ*
7'1"	36'	**45'**	132	1993	Longview, Huntington Park *RVP*

MAPLE (see also **Japanese Maples**, pages 70-71; and **Stripebark Maples**, page 72)

Circumference	Height	Crown Spread	AFA Points	Date Last Measured	Location and Nominators
Armstrong					*Acer* x *freemanii* 'Armstrong'
5'0"	69'	25'	**135**	1990	Tacoma, Mountain View Cemetery *RVP*
4'4"	**71'**	21'	128	1990	Tacoma, Mountain View Cemetery *RVP*
Black					*Acer saccharum* ssp. *nigrum*
11'1"	79'	**76'**	231	1992	Sumner, 315 Sumner Ave *KVP, RVP*
Coliseum					*Acer cappadocicum*
10'8"	62'	**57'**	204	1987	Tacoma, Pt Defiance Park *ALJ, RVP* (*photo page 68*)
7'10"	**91'**	51'	198	1987	Tacoma, Wright Park *RVP*
Columnar Red					*Acer rubrum* 'Columnare'
3'2"	66'	24'	**110**	1993	Seattle, Washington Park Arboretum, 32N 3W *ALJ, RVP*
2'11"	**69'**	21'	109	1992	Seattle, Washington Park Arboretum, 32N 3W *ALJ, RVP*

67

INTRODUCED TREES

Circumference	Height	Crown Spread	AFA Points	Date Last Measured	Location and Nominators
Cretan					*Acer sempervirens*
2'0"	25'	23'	55	1995	Seattle, Carl S English Gardens, Ballard Locks *RVP*
Crimson King					*Acer platanoides* 'Crimson King'
8'4"	54'	45'	**165**	1995	Brinnon, Whitney Gardens *RGB, RVP*
6'4"	52'	**51'**	141	1993	Steilacoom, Western Washington State Hospital *RVP*
Cutleaf Silver					*Acer saccharinum* 'Wieri Laciniatum'
8'5"	81'	**79'**	202	1993	Tacoma, Old Tacoma Cemetery *RVP*
English Field					*Acer campestre*
11'2"	83'	64'	233	1987	Tacoma, Wright Park *ALJ, RVP* (photo below)
8'3"	55'	**74'**	172	1992	Tacoma, Point Defiance Park, W end of rose garden *RVP*
Erect Norway					*Acer platanoides* 'Erectum'
8'1"	64'	47'	173	1992	Seattle, Washington Park Arboretum, 1S 2E *ALJ, RVP*
Faassen's Black					*Acer platanoides* 'Faassen's Black'
6'8"	40'	**41'**	130	1993	Mount Vernon, WSU Experiment Station *RGB, ALJ*
2'4"	**59'**	31'	95	1993	Seattle, Kubota Gardens Park *ALJ*
Kimball					*Acer macrophyllum* 'Kimballiae'
2'10"	31'	**48'**	77	1995	Seattle, Washington Park Arboretum, 9N 7E *KVP, RVP*
Lobel's					*Acer cappadocicum* ssp. *lobelii*
4'7"	57'	31'	120	1993	Seattle, Washington Park Arboretum, 10N 4W *RVP*
Miyabe					*Acer miyabei*
3'6"	36'	**38'**	87	1993	Spokane, Finch Arboretum *ALJ*
Montpelier					*Acer monspessulanum*
7'11"	42'	46'	**148**	1992	Bellingham, 405 Fieldstone Rd *RGB*
4'7"	**57'**	43'	123	1988	Tacoma, Wright Park *RVP*
6'4"	46'	**54'**	135	1992	Tacoma, Point Defiance Park *RVP*

This rugged tree is the state's largest **Coliseum Maple** which is located in Point Defiance Park in Tacoma. A beautiful but seldom planted tree, this maple is native to temperate southern Asia, from Turkey to China.

The **English Field Maple** is an often shrubby tree with tiny leaves and a twiggy habit that fool many into thinking it not a maple, yet its 'helicopter' seeds give it away. This outsized specimen in Tacoma's Wright Park is nearly as big as any in its native England.

Circumference	Height	Crown Spread	AFA Points	Date Last Measured	Location and Nominators
Newton Sentry					*Acer saccharum* **'Newton Sentry'**
2'5"	**60'**	12'	**92**	1988	Spokane, Finch Arboretum *ALJ, RVP*
Norway					*Acer platanoides*
13'1"	73'	76'	**249**	1992	Tacoma, 4620 Vickery Ave E *RGB*
8'11"	**97'**	72'	222	1988	Walla Walla, Pioneer Park *RVP*
9'5"	78'	**91'**	214	1993	Seattle, Volunteer Park, N of museum *RVP*
Red					*Acer rubrum*
10'11"	73'	75'	**223**	1988	Wenatchee, 101 Garfield *RVP*
9'1"	86'	**79'**	215	1987	Tacoma, Wright Park *RVP*
7'9"	**92'**	73'	203	1988	Tacoma, Wright Park *RVP*
Schwedler					*Acer platanoides* **'Schwedleri'**
8'3"	**88'**	49'	**199**	1987	Tacoma, Wright Park *RVP*
9'2"	73'	**56'**	197	1987	Seattle, Me-Kwa-Mooks Park *ALJ*
Seattle Sentinel					*Acer macrophyllum* **'Seattle Sentinel'**
6'6"	60'	22'	**143**	1988	Seattle, 18th Ave E & E Madison St *ALJ*
4'4"	**80'**	27'	139	1988	Seattle, Washington Park Arboretum, 29N 4E *ALJ*
Silver					*Acer saccharinum*
22'6"	84'	84'	**375**	1993	College Place, 725 SE Elm St *Shirley Muse*
20'0"	101'	100'	366	1988	Walla Walla, Whitman College, 364 Boyer *RVP (photo below)*
15'1"	**127'**	99'	333	1988	Walla Walla, Pioneer Park *RVP*
13'6"	98'	**119'**	287	1988	Spokane, Manito Park *ALJ, RVP*
Sugar					*Acer saccharum*
11'3"	86'	**83'**	**242**	1993	Tacoma, Wright Park *RVP (photo below)*
11'0"	90'	72'	**240**	1993	Seattle, 615 17th Ave E *ALJ*
12'2"	63'	65'	225	1995	Bellingham, Laurel Park *RGB*
9'5"	**93'**	79'	226	1990	Walla Walla, 343 Catherine *Shirley Muse*

The **Silver Maple** is well known to folks throughout the Midwest, where it is used for windbreaks, but its rapid growth and invasive roots make it a problem in many cities. If given enough room, however, it can grow into an impressive tree, such as this one in Walla Walla.

Certainly one of our country's most popular trees, the **Sugar Maple** has durable wood, delicious syrup, and unbeatable fall colors. This giant, in Tacoma's Wright Park, is among our largest and probably the most perfectly formed example.

INTRODUCED TREES

Japanese Maples

Japanese Maples are very popular ornamental trees in the Pacific Northwest, particularly west of the mountains, where cold temperatures are less common. Although most people think of *Acer palmatum* and its multitude of cultivars as the 'true' Japanese Maples, there are many other small-growing Japanese and East Asian maple species that are highly ornamental, hence their inclusion with Japanese Maples within the nursery trade. Below is a sampling of the larger and older specimens in Washington.

Circumference	Height	Crown Spread	AFA Points	Date Last Measured	Location and Nominators
Amur					***Acer tataricum* ssp. *ginnala***
5'5"	24'	**45'**	**100**	1993	Seattle, 4260 NE 113th St at Sand Point Way *ALJ, RVP*
3'0"	**33'**	43'	80	1988	Spokane, Finch Arboretum *ALJ, RVP*
Bloodgood					***Acer palmatum* 'Bloodgood'**
3'0"	**25'**	**25'**	67	1995	Seattle, 4835 Fauntleroy Way SW *RGB*
Burgundy Lace					***Acer palmatum* 'Burgundy Lace'**
3'4"	24'	24'	70	1995	Brinnon, Whitney Gardens *RGB, RVP*
Coral Bark					***Acer palmatum* 'Sango Kaku'**
2'2"	27'	25'	59	1993	Tacoma, Clover Park Voc Tech, E of Arboretum *RGB, RVP*
Fern Leaf					***Acer japonicum* 'Aconitifolium'**
1'10" 1 of 3 stems	**29'**	27'	58	1993	Richmond Beach, 714 NW 189th Ln *RGB, RVP*
1'11"	20'	19'	48	1993	Tacoma, Pt. Defiance Park, Japanese Garden *RVP*
1'8"	24'	**29'**	51	1993	Tacoma, Old Tacoma Cemetery *RVP*
Golden Fullmoon					***Acer shirasawanum* 'Aureum'**
1'6" best of 2 stems	**29'**	**18'**	51	1993	Tacoma, 676 N D St *Dick North*
Japanese					***Acer palmatum***
8'1"	**42'**	46'	**150**	1988	Seattle, 29th Ave W & W Crockett St *ALJ*
4'3"	39'	**51'**	103	1993	Lakewood, Lakewold Gardens *RGB, RVP*
Japanese Red					***Acer palmatum* 'Atropurpureum'**
6'11"	30'	**49'**	**125**	1990	Tacoma, 2920 N 15th *KVP, RVP*
4'9"	**39'**	45'	107	1990	Lakewood, Lakewold Gardens *KVP, RVP*
Matsu Kaze					***Acer palmatum* 'Matsu Kaze'**
2'0"	**31'**	**25'**	61	1995	Seattle, Washington Park Arboretum, 32N 1E *RVP*
Osakazuki					***Acer palmatum* 'Osakazuki'**
5'2"	**37'**	**40'**	**109**	1995	Seattle, 2117 E Shelby St *RVP*
4'11"	35'	41'	104	1995	Seattle, 2117 E Shelby St *RVP*
Painted					***Acer pictum***
2'8"	**63'**	28'	**102**	1993	Seattle, Washington Park Arboretum, 20N, 1W *ALJ*
4'11"	30'	**35'**	98	1990	Tacoma, Jefferson Park *KVP, RVP*
Paperbark					***Acer griseum***
4'1"	31'	24'	86	1988	Seattle, Washelli Cemetery *ALJ*
3'1"	**38'**	**31'**	83	1988	Seattle, Washington Park Arboretum, 26N, 0 *ALJ*
Pink Variegated Japanese					***Acer palmatum* 'Kagiri Nishiki'**
1'9"	**26'**	**24'**	53	1995	Seattle, Washington Park Arboretum, 32N 3E *RVP*

Circumference	Height	Crown Spread	AFA Points	Date Last Measured	Location and Nominators

Red Lace Leaf Japanese — *Acer palmatum* var. *dissectum* f. *atropurpureum*

Circumference	Height	Crown Spread	AFA Points	Date Last Measured	Location and Nominators
3'6"	10'	17'	56	1993	Gig Harbor, 4713 24th Ave NW *Dick North*
1'10"	**13'**	**21'**	40	1993	Tacoma, 518 N E St *KVP, RVP*

Red Nail Japanese — *Acer palmatum* 'Tsuma Beni'

Circumference	Height	Crown Spread	AFA Points	Date Last Measured	Location and Nominators
2'1"	32'	23'	63	1995	Seattle, Washington Park Arboretum, 32N 2E *RVP*

Seven-lobed Japanese — *Acer palmatum* ssp. *amoenum*

Circumference	Height	Crown Spread	AFA Points	Date Last Measured	Location and Nominators
5'4"	27'	**37'**	100	1993	Tacoma, 3922 N Mason *KVP, RVP*
2'8"	**34'**	27'	73	1995	Seattle, Washington Park Arboretum, 33N, 1E *RVP*

Shantung — *Acer truncatum*

Circumference	Height	Crown Spread	AFA Points	Date Last Measured	Location and Nominators
4'0"	29'	49'	89	1993	Seattle, Washington Park Arboretum, 12N, 3W *RVP*

Siebold — *Acer sieboldianum*

Circumference	Height	Crown Spread	AFA Points	Date Last Measured	Location and Nominators
2'10"	26'	23'	66	1995	Seattle, Washington Park Arboretum, 32N, 1E *RVP*

Trident — *Acer buergerianum*

Circumference	Height	Crown Spread	AFA Points	Date Last Measured	Location and Nominators
2'1"	44'	23'	75	1990	Seattle, Washington Park Arboretum, 11N, 3E *ALJ*
3'0"	31'	27'	74	1990	Seattle, 727 Bellevue Ave E *ALJ*

Vineleaf — *Acer cissifolium*

Circumference	Height	Crown Spread	AFA Points	Date Last Measured	Location and Nominators
3'1"	39'	41'	86	1993	Seattle, Washington Park Arboretum, 12N, 4E *RVP*

Whitney Red — *Acer palmatum* 'Whitney Red'

Circumference	Height	Crown Spread	AFA Points	Date Last Measured	Location and Nominators
4'3"	31'	33'	90	1995	Brinnon, Whitney Gardens *RGB, RVP*

Wou Nishiki — *Acer palmatum* 'Wou Nishiki'

Circumference	Height	Crown Spread	AFA Points	Date Last Measured	Location and Nominators
2'5"	36'	27'	72	1995	Seattle, Washington Park Arboretum, 32N 3E *RVP*

Popular throughout western Washington is the purple-leaved **Japanese Red Maple.** A large-growing form of the Japanese Maple, this cultivar has purple new leaves, red seeds, and excellent fall colors. This specimen, in a Tacoma yard, is our biggest.

INTRODUCED TREES

Stripebark Maples

Also called Snakebark Maples, these small, forest understory trees largely come from East Asia. The exception, Moosewood, is the only Stripebark native to North America. Mature trees such as the State Champions develop very attractive branch and bark patterns in addition to excellent fall colors.

As can be seen from the listing below, Washington Park Arboretum in Seattle is our best place to view Stripebark Maples. The Stripebarks tend to favor a woodland setting, and all but one of the trees listed occur in a lightly forested environment. The exception is Huntington Park in Longview, which contains several other State Champion trees.

The **Manchurian Stripebark**, pictured, shows its characteristic smooth bark with distinctive striping patterns developed on the trunk and branches. Many of our largest Stripebarks can be seen along Arboretum Drive in Seattle.

Circumference	Height	Crown Spread	AFA Points	Date Last Measured	Location and Nominators
Chinese Stripebark					*Acer tetramerum* var. *lobatum*
2'4"	43'	33'	79	1995	Seattle, Washington Park Arboretum, 14N, 9E *RVP*
Gray-Budded Snakebark					*Acer rufinerve*
3'4"	49'	44'	100	1990	Seattle, Washington Park Arboretum, 19N, 0 *ALJ*
1'11"	62'	23'	91	1990	Seattle, Washington Park Arboretum, 20N, 0 *RVP*
Hawthorn-Leaved Snakebark					*Acer crataegifolium*
2'3"	21'	23'	54	1995	Seattle, Washington Park Arboretum, 32N, 1E *RVP*
Hers's Stripebark					*Acer davidii* 'Hersii'
4'10"	36'	41'	104	1993	Longview, Huntington Park *RVP*
4'0"	46'	41'	104	1993	Seattle, Washington Park Arboretum, 16N, 7E *ALJ*
3'1"	56'	23'	99	1993	Seattle, Washington Park Arboretum, 20N, 0 *ALJ*
Manchurian Stripebark					*Acer tegmentosum*
3'4"	46'	36'	95	1990	Seattle, Washington Park Arboretum, 15N, 7E *ALJ*
Moosewood					*Acer pensylvanicum*
2'0"	21'	16'	49	1993	Seattle, Volunteer Park *ALJ*
1'3"	26'	21'	46	1993	Dayton, Dayton City Park *Shirley Muse, RGB, RVP*
Père David's Stripebark					*Acer davidii*
3'4"	57'	43'	108	1993	Richmond Beach, 20066 15th Ave NW *ALJ*
3'9"	33'	43'	89	1990	Seattle, Washington Park Arboretum, 15N, 8E *ALJ*
Red Snakebark					*Acer capillipes*
3'3"	56'	33'	103	1990	Seattle, Washington Park Arboretum, 20N, 0 *ALJ, RVP*
4'1"	38'	43'	98	1992	Seattle, Washington Park Arboretum, 2S, 2E *ALJ*
Red-Twigged Moosewood					*Acer pensylvanicum* 'Erythrocladum'
1'6"	26'	13'	47	1995	Federal Way, Rhododendron Spicies Foundation *RGB*

Circumference	Height	Crown Spread	AFA Points	Date Last Measured	Location and Nominators
Sweet Shadow					*Acer saccharum* 'Sweet Shadow'
5'0"	44'	**39'**	**114**	1995	Renton, 300 SW 7th *RGB*
2'9"	**48'**	35'	89	1993	Bothell, Rhody Ridge Arboretum *RGB*
Sycamore					*Acer pseudoplatanus*
13'11"	80'	**85'**	**268**	1992	Lyman, across from 110 E First St *RGB*
6'4"	**89'**	48'	177	1988	Seattle, Seattle Pacific University *ALJ, RVP*
Variegated Sycamore					*Acer pseudoplatanus* **f. variegatum**
4'7"	**52'**	**40'**	**117**	1990	Seattle, Mt Pleasant Cemetery *ALJ*
4'9"	39'	39'	107	1995	Everett, 2701 17th St *RGB*
Wineleaf Sycamore					*Acer pseudoplatanus* 'Atropurpureum'
12'9" below branching	71'	59'	**239**	1992	Carnation, Tolt Middle School *RGB*
6'6"	**82'**	47'	172	1990	Tacoma, Wright Park *RVP*
10'7"	54'	**82'**	201	1992	Sedro Woolley, 2043 F & S Grade Rd *RGB*

MAYTEN
Maytenus boaria

Circumference	Height	Crown Spread	AFA Points	Date Last Measured	Location and Nominators
2'5"	**38'**	**27'**	**74**	1995	Seattle, Washington Park Arboretum, 11N 6E *RGB*

MEDLAR
Mespilus germanica

Circumference	Height	Crown Spread	AFA Points	Date Last Measured	Location and Nominators
2'5"	12'	**27'**	**48**	1993	Seattle, Washington Park Arboretum, 11N 3W *RVP*
2'2"	**13'**	25'	**45**	1990	Bainbridge Island, Restoration Point *ALJ*

Certainly one of our most distinctive trees, the **Monkey-Puzzle** is a bizarre conifer from Chile. The tree pictured grows near the Skagit River Delta, and is among our largest.

The **European Mountain-Ash** is perfectly at home in our climate. This National Champion at Woodland Park Zoo in Seattle has grown to proportions that exceed those in its native Europe.

INTRODUCED TREES

Circumference	Height	Crown Spread	AFA Points	Date Last Measured	Location and Nominators

MONKEY-PUZZLE
Araucaria araucana

10'2"	66'	43'	**199**	1993	Aberdeen, Fern Hill Cemetery *RGB*
9'3"	69'	43'	**191**	1993	Skagit City, 1954 Skagit City Rd *RVP (photo page 73)*
8'10"	**72'**	33'	186	1988	Bremerton, NE corner of 3rd & Naval *RVP*

MOUNTAIN-ASH (see also SERVICE TREE, page 89; and WHITEBEAM, page 97)

Chinese — *Sorbus scalaris*

3'10"	59'	47'	117	1995	Seattle, Washington Park Arboretum, 24N 4E *KVP, RVP*

Columnar — *Sorbus aucuparia* **f.** *fastigiata*

6'6"	48'	23'	132	1992	Seattle, 371 NW 48th St *RVP*

European — *Sorbus aucuparia*

11'0"	43'	41'	❖185	1988	Seattle, Woodland Park Zoo *ALJ (photo page 73)*
6'3"	**85'**	43'	171	1995	Bremerton, Naval Shipyard, Bd 644 *Louise Reh, KVP RVP*

Hupeh — *Sorbus hupehensis*

3'8"	49'	28'	100	1993	Seattle, Ayre, The Highlands *RGB*

Japanese — *Sorbus commixta*

2'5"	52'	23'	87	1995	Seattle, Washington Park Arboretum, 26N 5E *KVP, RVP*
3'0"	34'	**29'**	77	1995	Seattle, Washington Park Arboretum, 25N 4E *KVP, RVP*

Pratt — *Sorbus prattii*

1'11"	26'	19'	54	1995	Lake Forest Park, 18540 26th Ave NE *Sallie Allen*

Weeping European — *Sorbus aucuparia* 'Pendula'

5'1"	17'	31'	86	1993	Lynnwood, Greenland Cemetery, Filbert Rd *RGB, ALJ*

MULBERRY

Black — *Morus nigra*

4'3"	27'	29'	85	1992	Vashon Island, 10325 SW Cemetery Rd *Mike Lee, RGB*

Weeping — *Morus alba* 'Pendula'

3'7"	13'	12'	59	1993	Lakewood, Veterans Hospital entrance *KVP, RVP*
3'2"	**16'**	**15'**	58	1995	Seattle, 8311 16th Ave NW *RVP*

White — *Morus alba*

10'10"	65'	51'	208	1993	Walla Walla, 1130 Howard St *RGB*

MYRTLE

Oregon — *Umbellularia californica*

15'7" below forking	53'	60'	255	1993	Vancouver, 3409 Main St, Pythian House *RVP*
11'7"	**71'**	**67'**	226	1990	Mercer Island, 9820 SE 35th Pl *RGB, RVP*

OAK (see also Eastern North American Oaks, pages 76-77)

Armenian — *Quercus pontica*

4'10"	21'	21'	84	1991	Bremerton, Olympic College *Jonathan Schwartz, RGB*

Bamboo-leaf — *Quercus myrsinifolia*

3'4"	44'	31'	92	1988	Redmond, Marymoor Park *ALJ, RVP*

California Black — *Quercus kelloggii*

7'8"	84'	75'	195	1993	Fall City, 4328 338th Place SE *RVP*

Canyon Live — *Quercus chrysolepis*

9'7"	47'	47'	174	1990	Big Lake, 1749 State Hwy 9 *RGB*
9'2"	46'	**60'**	171	1990	Big Lake, 1749 State Hwy 9 *RGB*
8'0"	**51'**	57'	161	1987	Seattle, Carl S. English Gardens at the Ballard Locks *ALJ*

Chinese Cork — *Quercus variabilis*

3'10"	85'	31'	139	1990	Seattle, Washington Park Arboretum, 43N 1E *ALJ, RVP*

Circumference	Height	Crown Spread	AFA Points	Date Last Measured	Location and Nominators
Coast Live					*Quercus agrifolia*
7'8"	**45'**	**43'**	**148**	1990	Seattle, 6250 Lake Shore Dr S *RVP*
6'3"	**45'**	41'	130	1989	Seattle, 6049 NE Keswick Dr *ALJ*
Cork					*Quercus suber*
6'4"	**40'**	36'	**125**	1993	Seattle, 2864 44th Ave W *RVP*
5'8"	38'	**39'**	116	1995	Seattle, University of Washington, HUB *ALJ*
Cypress					*Quercus robur* 'Fastigiata'
10'3"	91'	**56'**	**228**	1987	Tacoma, C St & N 5th St *RVP*
7'1"	**97'**	33'	190	1987	Tacoma, C St & N 5th St *RVP*
Daimyo					*Quercus dentata*
8'2"	70'	**61'**	183	1988	Seattle, Green Lake Park *ALJ*
Durmast					*Quercus petraea*
10'4"	86'	**79'**	230	1994	Seattle, Garfield High School playground *ALJ*
English					*Quercus robur*
14'10"	102'	89'	❖302	1993	Olympia, Capitol grounds *RVP* *(photo below)*
11'11"	102'	82'	265	1993	Bellingham, Elizabeth Park *RVP*
9'8"	93'	**103'**	235	1992	Tacoma, Wright Park, near wading pool *RVP*
Holm					*Quercus ilex*
7'0"	47'	**55'**	145	1989	Bainbridge Island, Lynnwood Center, Beck & Blakeley Sts *ALJ*
Huckleberry					*Quercus vaccinifolia*
3'8"	27'	25'	77	1995	Seattle, Carl S English Gardens, Ballard Locks *KVP, RVP*
Interior Live					*Quercus wislizenii*
7'4"	62'	69'	167	1987	Seattle, Washington Park Arboretum, 37N 4E *ALJ*
Sawtooth					*Quercus acutissima*
2'11"	69'	37'	113	1993	Seattle, Washington Park Arboretum, 40N 2W *ALJ*

The **English Oak** pictured here is on the lawn at the State Capitol in Olympia. In 1987, when I first measured it, its AFA point total was 279 – 9 short of the National Champion. This year I remeasured it – the trunk had expanded from 13'5" to 14'10", totaling 302 points! A new record!

The **Pin Oak**, a riverine tree from the eastern United States, is very commonly planted in our state. Our largest specimen, shown here, may be found in historic Tacoma. The tree was planted by E.R. Roberts, the person responsible for planting beautiful Wright Park, also in Tacoma.

INTRODUCED TREES

Eastern North American Oaks

Oaks have been popular ornamental trees in Washington since the first white settlers arrived over 100 years ago. These noble trees love our climate and can grow to great size. Since Washington has only one native oak species, it is understandable that when folks travel from the East, where there are dozens of native species, they should plant oaks in their parks and gardens. Today we have a rich collection of oak species, many of which have developed impressive proportions. The Red Oak is the largest of all, and at 132 feet, is the second tallest species of introduced broadleaf tree in the state.

Circumference	Height	Crown Spread	AFA Points	Date Last Measured	Location and Nominators
Black					**Quercus velutina**
11'9"	95'	**86'**	**258**	1988	Sedro Woolley, 329 Talcott *RVP*
10'8"	**99'**	74'	245	1988	Sedro Woolley, 329 Talcott *RVP*
Bur					**Quercus macrocarpa**
13'7"	**101'**	103'	**290**	1988	Walla Walla, 42 E Maple St *RVP*
13'4"	100'	**107'**	287	1988	Walla Walla, Whitman College, 364 Boyer *RVP*
Cherrybark					**Quercus falcata** var. **pagodifolia**
4'10"	74'	**59'**	**147**	1993	Seattle, Washington Park Arboretum, 53N 13E *RVP*
4'4"	**79'**	49'	143	1993	Seattle, Washington Park Arboretum, 53N 13E *RVP*
Chestnut					**Quercus prinus**
12'7"	**104'**	**95'**	**276**	1990	Seattle, 627 36th Ave E *ALJ, RVP*
Chinquapin					**Quercus muehlenbergii**
8'7"	56'	**49'**	**171**	1995	Seattle, Loyal Heights Playfield *ALJ*
4'6"	**73'**	40'	139	1989	Seattle, Woodland Park Zoo, Asian Tropical Forest *ALJ*
Laurel					**Quercus laurifolia**
4'7"	**68'**	**57'**	**137**	1995	Seattle, Washington Park Arboretum, 44N 1E *RGB*
Northern Pin					**Quercus ellipsoidalis**
10'4"	**77'**	**77'**	**220**	1993	Mt Vernon, Kincaid St at I-5 exit ramp *RVP*
Pin					**Quercus palustris**
12'10"	**114'**	80'	**288**	1995	Tacoma, 601 N Yakima *RVP (photo page 75)*
12'7"	94'	**93'**	268	1995	North Bend, W of North Bend Elem School *RGB*
Red					**Quercus rubra**
15'2"	**132'**	93'	**337**	1990	Grays River, Swede Park *Bob Pyle*
15'4"	107'	117'	320	1990	Tacoma, Wright Park *ALJ (photo page 77)*
15'7"	95'	88'	304	1990	Grays River, Swede Park *Bob Pyle*
13'5"	85'	**124'**	272	1990	Tacoma, Wright Park *RVP*
Scarlet					**Quercus coccinea**
11'9"	**117'**	87'	**280**	1993	Walla Walla, Pioneer Park *RVP*
14'4"	74'	**101'**	271	1993	Longview, Washington Way & 21st Ave *RVP (photo page 101)*
12'7"	100'	80'	271	1992	Sumner, 1228 Main St, Ryan House-museum *RGB, RVP*
Shingle					**Quercus imbricaria**
9'6"	**103'**	64'	**233**	1988	Seattle, 1030 39th Ave E *RVP*
7'7"	92'	**74'**	201	1987	Seattle, University of Washington, Parrington Lawn *ALJ (photo page 77)*

Circumference	Height	Crown Spread	AFA Points	Date Last Measured	Location and Nominators
Shumard					*Quercus shumardii*
7'3"	65'	48'	**164**	1993	Camas, 800 NE 3rd Ave, Safeway *RGB, RVP*
5'10"	**77'**	**55'**	161	1993	Seattle, Washington Park Arboretum, 50N, 2E *ALJ*
Southern Red					*Quercus falcata*
6'5"	**95'**	48'	**184**	1988	Seattle, Pacific Medical Center *ALJ*
7'5"	74'	**54'**	176	1988	Seattle, Pacific Medical Center *ALJ*
Swamp Chestnut					*Quercus michauxii*
5'8"	**83'**	59'	**166**	1988	Seattle, Washington Park Arboretum, 41N 2E *RVP*
6'2"	74'	59'	163	1988	Seattle, Washington Park Arboretum, 41N 2E *RVP*
Swamp White					*Quercus bicolor*
8'4"	**79'**	44'	**190**	1995	Bellingham, Western Washington University, across from bookstore *RGB*
6'1"	64'	**55'**	151	1988	Seattle, Green Lake Park *ALJ*
Water					*Quercus nigra*
7'11"	**76'**	**68'**	**188**	1995	Seattle, Washington Park Arboretum, 45N 1E *ALJ*
8'1"	63'	59'	175	1995	Seattle, University of Washington, Architecture Hall *ALJ*
White					*Quercus alba*
14'2"	81'	**102'**	**276**	1988	Yakima, Stone Church, S 17th & W Chestnut Sts *RVP*
11'1"	**94'**	80'	247	1988	Seattle, 1809 Parkside Dr E *ALJ*
Willow					*Quercus phellos*
7'2"	73'	72'	177	1988	Seattle, Lake Washington Blvd & 36th Ave E *ALJ*
6'4"	**84'**	64'	**176**	1993	Seattle, Washington Park Arboretum, 38N 1W *RVP*

The **Red Oak** is the largest species of Eastern oak in Washington. The specimen pictured here (left) in Wright Park in Tacoma is slightly over 100 years old and already over five feet in diameter. It is a co-champion for size with trees in Grays River.

Shingle Oak is an Eastern oak with unoaklike leaves. Instead of having teeth, the leaves have smooth edges. The lovely tree pictured (right) is on the University of Washington campus.

INTRODUCED TREES

Circumference	Height	Crown Spread	AFA Points	Date Last Measured	Location and Nominators
Silverleaf					*Quercus hypoleucoides*
4'10"	**42'**	**35'**	**109**	1992	Seattle, Carl S English Gardens, Ballard Locks *ALJ*
Turkish					*Quercus cerris*
10'6"	86'	**72'**	**230**	1990	Woodinville, Chateau Ste Michelle Winery *RVP*
5'9"	**111'**	44'	186	1994	Seattle, Mt Baker Park *ALJ*
Ubame					*Quercus phillyraeoides*
4'10"	**34'**	**35'**	**101**	1992	Seattle, Carl S English Gardens, Ballard Locks *ALJ*
Valley					*Quercus lobata*
12'3"	**78'**	**76'**	**244**	1987	Seattle, Seattle Pacific University *ALJ*

OSAGE ORANGE
Maclura pomifera

Circumference	Height	Crown Spread	AFA Points	Date Last Measured	Location and Nominators
6'8"	56'	**53'**	**149**	1993	Puyallup, 62nd Ave E at River Rd *RGB*
6'6"	56'	43'	**145**	1993	Walla Walla, 604 W Alder St *Shirley Muse*
largest of 2 trunks					
4'10"	**70'**	33'	136	1987	Seattle, Mt Baker Blvd *ALJ*

OSMANTHUS

Circumference	Height	Crown Spread	AFA Points	Date Last Measured	Location and Nominators
Hybrid					*Osmanthus* x *fortunei*
5'6"	**24'**	**24'**	**96**	1992	Seattle, 4115 Brooklyn Ave NE *ALJ*
Japanese					*Osmanthus heterophyllus*
3'11"	18'	**33'**	**73**	1993	Tacoma, Pt Defiance Park *RVP*
3'1"	**22'**	27'	66	1993	Seattle, Washington Park Arboretum, 39N 3W *ALJ*
1 of 3 trunks					

The **Chinese Windmill Palm**, our only tree-sized, hardy palm, is occasionally found in western Washington. As can be seen from the photo, our champion is in a protected alcove of an apartment complex in Tacoma. This 38' tree's closest competitor is only 27' tall.

The **Common Pear** is frequent around farms and old homesteads throughout Washington. A pioneer happened to plant one in the rich soil near Walla Walla about a hundred years ago. The result – the largest pear tree in the nation.

Circumference	Height	Crown Spread	AFA Points	Date Last Measured	Location and Nominators

PAGODA TREE

Japanese — _Sophora japonica_

| 8'7" | **73'** | **60'** | **191** | 1990 | Seattle, 7140 55th Ave S *ALJ, RVP* |
| 9'3" | 53' | 59' | 179 | 1987 | Seattle, Broadway reservoir *ALJ* |

Weeping — _Sophora japonica_ 'Pendula'

| 4'8" | **32'** | **39'** | **98** | 1995 | Seattle, Washington Park Arboretum, 18N 6E *RVP* |

PALM

Chinese Windmill — _Trachycarpus fortunei_

| 1'9" | **38'** | **11'** | **62** | 1992 | Tacoma, Bay View Apts. *KVP, RVP* (photo page 78) |

PAWPAW

Asimina triloba

| 2'10" | **30'** | 12' | **67** | 1990 | Walla Walla, 620 S Howard *Shirley Muse* |
| 2'8" | 18' | **21'** | 55 | 1992 | Vashon Island, 10325 SW Cemetery Rd *Mike Lee* |

PEAR

Asian — _Pyrus pyrifolia_ var. _culta_

| 4'10" | **40'** | **47'** | **110** | 1993 | Mountlake Terrace, 5302 212 St SW *RGB* |

Callery — _Pyrus calleryana_

| 3'10" | **48'** | 25' | **100** | 1995 | Vancouver, 1200 Ft Vancouver Way, Clark Public Utilities *RGB* |
| 3'2" | 42' | **33'** | 88 | 1992 | Seattle, Washington Park Arboretum, 23N 4W *ALJ* |

Common — _Pyrus_ x _communis_

| 14'6" | 59' | **56'** | ❖**247** | 1990 | Lowden, US 12, mm 327 *S Muse, M Drawson, RVP* (photo page 78) |
| 5'2" | **67'** | 35' | 138 | 1987 | Seattle, Jefferson Park Golf Course *ALJ* |

Hybrid Snow — _Pyrus_ x _salvifolia_

| 3'10" | **50'** | 27' | **103** | 1995 | Seattle, Rainier Playfield *RGB* |

Willow-Leaved Hybrid — _Pyrus salicifolia_ hybrid

| 2'8" | **35'** | **37'** | **76** | 1993 | Seattle, Washington Park Arboretum, 18N 6W *RVP* |

PECAN

Carya illinoinensis

| 10'11" | **120'** | **76'** | **270** | 1993 | Walla Walla, 384 S Palouse *RVP* |

PERSIMMON

American — _Diospyros virginiana_

3'6"	40'	29'	**89**	1992	Steilacoom, 1614 Rainier St *RGB, RVP*
3'3"	35'	**32'**	82	1989	Seattle, Denny Park *ALJ*
2'0"	**43'**	20'	72	1989	Seattle, Washington Park Arboretum, 12N 1W *ALJ*

Date-plum — _Diospyros lotus_

| 4'5" | 40' | **39'** | **103** | 1990 | Seattle, Kinnear Park *ALJ* |
| 1'7" | **45'** | 23' | 70 | 1990 | Seattle, Carl S English Gardens, Ballard Locks *ALJ* |

Japanese — _Diospyros kaki_

| 2'9" | **35'** | **25'** | **74** | 1995 | Seattle, 10th Ave S at S Thistle St *ALJ* |

PHELLODENDRON

Amur Cork Tree — _Phellodendron amurense_

| 5'7" | **41'** | **54'** | **121** | 1993 | Seattle, 3414 Shore Dr, Broadmoor *RVP* |

PHOTINIA

Birmingham — _Photinia fraseri_ 'Birmingham'

| 3'10" | **26'** | **29'** | **79** | 1995 | Seattle, Washington Park Arboretum, 16N 4E *RGB* |

INTRODUCED TREES

Circumference	Height	Crown Spread	AFA Points	Date Last Measured	Location and Nominators

Chinese   *Photinia serratifolia*

Circumference	Height	Crown Spread	AFA Points	Date Last Measured	Location and Nominators
4'1" largest of 2 stems	48'	36'	106	1993	Seattle, 8912 Fauntleroy Way SW *Mike Lee*

Fraser   *Photinia fraseri*

Circumference	Height	Crown Spread	AFA Points	Date Last Measured	Location and Nominators
3'0"	27'	24'	69	1995	Bellevue, 417 Bellevue Way ·*RGB*

Variegated   *Photinia serratifolia* 'Nova Lineata'

Circumference	Height	Crown Spread	AFA Points	Date Last Measured	Location and Nominators
3'5"	26'	32'	75	1995	Seattle, 9121 Fauntleroy Way SW *RGB*

PINE

Austrian   *Pinus nigra*

Circumference	Height	Crown Spread	AFA Points	Date Last Measured	Location and Nominators
11'3"	90'	61'	240	1993	Seattle, Washelli Cemetery *ALJ*
8'9"	66'	63'	189	1992	Tacoma, Point Defiance Park *RVP*

Bishop   *Pinus muricata*

Circumference	Height	Crown Spread	AFA Points	Date Last Measured	Location and Nominators
10'10"	76'	47'	218	1995	Chehalis, 2042 Bishop Rd *RGB*

Bristlecone   *Pinus aristata*

Circumference	Height	Crown Spread	AFA Points	Date Last Measured	Location and Nominators
3'3"	29'	30'	75	1988	Spokane, Finch Arboretum *ALJ, RVP*

Chinese   *Pinus tabulaeformis*

Circumference	Height	Crown Spread	AFA Points	Date Last Measured	Location and Nominators
8'9"	49'	56'	168	1995	Bellingham, NW corner of Potter & Moore Sts *RGB*
2'7"	69'	25'	106	1990	Wind River, USFS Arboretum *RVP*

Chinese White   *Pinus armandii*

Circumference	Height	Crown Spread	AFA Points	Date Last Measured	Location and Nominators
3'3"	63'	33'	110	1993	Seattle, Washington Park Arboretum, 54N 11E *RVP*

Pinus nigra is a variable European pine that grows in the Alps, Ukraine, and Mediterranean area. Those from the latter area are called **Corsican Pines**, and are characterized by strong growth and an open branch structure. Tacoma's Lincoln Park has the largest known *Pinus nigra* in the country.

The **Digger Pine** is also called Foothills Pine, as it is native to the foothills of California's Central Valley. This tree at the University of Washington has the multiple braching pattern characteristic of this species. This tree is no longer the Washington Champion, but is easily viewed.

INTRODUCED TREES

Circumference	Height	Crown Spread	AFA Points	Date Last Measured	Location and Nominators
Columnar Scots					*Pinus sylvestis* 'Fastigiata'
1'7"	**30'**	8'	**51**	1993	Seattle, Washington Park Arboretum, 32N 6E *ALJ*
Columnar White					*Pinus strobus* 'Fastigiata'
4'11"	**68'**	29'	**134**	1993	Milton, Greenwood Residences, 6th Ave *KVP, RVP*
Corsican					*Pinus nigra* var. *corsicana*
10'9"	**114'**	49'	❖**255**	1989	Tacoma, Lincoln Park *RVP (photo page 80)*
Coulter					*Pinus coulteri*
9'10"	88'	48'	**218**	1995	Seattle, University of Washington, Winkenwerder Hall *ALJ*
10'7"	70'	60'	**212**	1995	Kirkland, NE 66th Lane *RGB*
9'9"	78'	60'	**210**	1995	Bellevue, 203 140th NE *RGB*
6'4"	**105'**	40'	189	1992	Seattle, University of Washington, W of McMahon Hall *ALJ*
Crimean					*Pinus nigra* var. *pallasiana*
12'6"	**61'**	51'	**224**	1988	Mt Vernon, 1863 Bradshaw Rd *RVP*
topped; was 72' tall					
Digger					*Pinus sabiniana*
12'1"	79'	**71'**	**242**	1993	Vancouver, N St, end of 3700 blk *RVP*
9'9"	**86'**	40'	213	1990	Seattle, 814 McGilvra Blvd E *ALJ, RVP*
Eastern White					*Pinus strobus*
15'0"	78'	60'	**273**	1992	Vashon Island, 16606 99th Ave SW *RGB, RVP*
below branching					
10'10"	**105'**	67'	252	1992	Fircrest, 544 Ramsdell *KVP, RVP*
Himalayan White					*Pinus wallichiana*
8'9"	77'	61'	**197**	1993	Seattle, S Holly and 57th Ave S *ALJ*
8'1"	86'	51'	**196**	1988	Spokane, Paulsen Mansion, E 245 13th Ave E *ALJ, RVP*
6'6"	**99'**	35'	186	1990	Tacoma, Wright Park *RVP*
Italian Stone					*Pinus pinea*
8'5"	50'	41'	**161**	1992	Seattle, Broadmoor Golf Course *RGB*
6'9"	**53'**	**49'**	146	1988	Seattle, Washington Park Arboretum, 32N 5E *ALJ*
Jack					*Pinus banksiana*
3'9"	58'	19'	**108**	1988	Seattle, Volunteer Park *ALJ*
2'9"	**63'**	20'	101	1988	Seattle, Washington Park Arboretum, Foster Island *ALJ*
Japanese Black					*Pinus thunbergii*
4'11"	**70'**	31'	**137**	1987	Seattle, Ravenna Park *ALJ*
Japanese Red					*Pinus densiflora*
5'11"	66'	35'	**146**	1993	Seattle, Washington Park Arboretum, 36N 4W *RVP*
7'6"	43'	43'	**144**	1987	Seattle, Rainier Vista housing project *ALJ*
4'0"	**85'**	19'	**138**	1995	Bainbridge Island, Bainbridge Gardens Nursery *RGB*
Japanese White					*Pinus parviflora*
6'7"	44'	**39'**	**133**	1993	Milton, 101 Milton Way *KVP, RVP*
2'11"	**59'**	16'	98	1988	Wind River, USFS Arboretum *RVP*
Japanese White (Blue Form)					*Pinus parviflora* 'Glauca'
8'3"	33'	**37'**	**141**	1989	Tacoma, 3306 N Union *RVP*
below branching					
4'3"	**43'**	**37'**	103	1992	Puyallup, 404 23rd Ave SE *Martha Robbins*
Jeffrey					*Pinus jeffreyi*
14'4"	92'	49'	**277**	1995	Avon, 1303 Avon-Allen Rd *RVP*
15'1"	80'	37'	**270**	1992	Sedro Woolley, 1301 3rd St *RGB*
13'7"	81'	**68'**	**261**	1988	Centralia, 1230 Main *RVP*
10'2"	**117'**	51'	252	1988	Wind River, USFS Arboretum *RVP (photo page 82)*
Knobcone					*Pinus attenuata*
10'5"	48'	40'	**183**	1993	Tacoma, 3405 N 34th *RVP*
below forking					
6'0"	**81'**	35'	162	1993	Seattle, Washington Park Arboretum, 36N 6W *ALJ*

INTRODUCED TREES

Circumference	Height	Crown Spread	AFA Points	Date Last Measured	Location and Nominators
Korean					*Pinus koraiensis*
4'5"	50'	29'	**110**	1993	Puyallup, 11215 Valley Ave E *RGB*
4'8"	33'	39'	99	1990	Tacoma, 4613 S Park St *RVP*
2'2"	**59'**	19'	90	1990	Wind River, USFS Arboretum *RVP*
Lacebark					*Pinus bungeana*
3'2"	35'	**27'**	80	1995	Bellingham, Broadway Park *RGB*
1 of 2 stems					
2'4"	**40'**	15'	72	1995	Bellingham, Broadway Park *RGB*
Limber					*Pinus flexilis*
4'8"	65'	29'	**128**	1993	Seattle, Washington Park Arboretum, 36N 6W *ALJ*
3'2"	**71'**	31'	117	1990	Wind River, USFS Arboretum *RVP*
Loblolly					*Pinus taeda*
8'1"	50'	43'	**158**	1992	Issaquah, 10010 238th Way SE *RGB, RVP*
6'9"	**63'**	29'	151	1992	Sedro Woolley, DNR Research Station *RGB*
Macedonian					*Pinus peuce*
3'11"	67'	28'	**121**	1990	Seattle, Washington Park Arboretum, 23N 1W *RVP*
Maritime					*Pinus pinaster*
10'9"	**88'**	**51'**	**230**	1987	Allen, Cook Rd near Hwy 11 *RVP (photo below)*
Mexican Pinyon					*Pinus cembroides*
2'7"	**25'**	19'	**61**	1987	Seattle, Jefferson Park golf course *ALJ*
Monterey					*Pinus radiata*
7'7"	**98'**	28'	**195**	1992	Wauna, 8020 St Hwy 302, W of Burley Lagoon bridge *RVP*

A close cousin of our Ponderosa Pine, **Jeffrey Pine** is native to the Sierra Nevada and Siskiyou Mountains of California and Oregon. The Wind River Arboretum near Carson is home to this lofty specimen, Washington's tallest introduced pine.

Along the Mediterranean coast of Monte Carlo one may find the lovely **Maritime Pine** clinging to the shoreline cliffs. One of my favorite pines, it usually has a beautiful maroon-colored bark and a graceful, curving trunk. Washington's largest grows on Skagit Valley soil near Allen.

INTRODUCED TREES

Circumference	Height	Crown Spread	AFA Points	Date Last Measured	Location and Nominators
Montezuma					*Pinus montezumae*
4'6"	**49'**	36'	**112**	1993	Seattle, Washington Park Arboretum, 32N 5E *RVP*
Mountain					*Pinus uncinata*
3'8"	**62'**	16'	**110**	1987	Seattle, Volunteer Park *ALJ*
4'5"	37'	33'	98	1993	Tacoma, Pt Defiance Park, Japanese Garden *RVP*
Mugo					*Pinus mugo*
9'0" below forking	43'	**49'**	**163**	1993	Toppenish, Park at N D St & Lincoln Ave *RGB, RVP*
4'4"	**55'**	26'	113	1992	Sumner, Loyalty Park *RGB, RVP*
Pitch					*Pinus rigida*
3'10"	61'	33'	**115**	1988	Seattle, Washington Park Arboretum, 36N 6W *ALJ*
3'7"	**65'**	24'	**114**	1990	Wind River, USFS Arboretum *RVP*
4'10"	45'	34'	**111**	1988	Seattle, Seward Park *ALJ*
Red					*Pinus resinosa*
4'5"	**79'**	29'	**139**	1990	Wind River, USFS Arboretum *RVP*
4'9"	71'	31'	**135**	1988	Wind River, USFS Arboretum *RVP*
Scots					*Pinus sylvestris*
10'4"	60'	**55'**	**198**	1995	Mt Vernon, 1546 McLean Rd *RGB*
9'6"	68'	49'	**194**	1987	Wenatchee, 1220 Orchard *RVP*
4'3"	**106'**	27'	164	1988	Monroe, State Reformatory *RVP*
Shortleaf					*Pinus echinata*
5'5"	**80'**	33'	**153**	1995	Bellingham, Yew St Parks Dept *RGB, RVP*
5'7"	67'	35'	143	1993	Washougal, 733 E St *RGB, RVP*
Southwestern White					*Pinus strobiformis*
5'3"	74'	37'	**146**	1992	Seattle, University of Washington, McCarty Hall, S end *ALJ*
3'10"	**82'**	31'	136	1990	Wind River, USFS Arboretum *RVP*
Sugar					*Pinus lambertiana*
7'4"	**96'**	35'	**193**	1988	Wind River, USFS Arboretum *RVP*
8'4"	78'	35'	187	1987	Seattle, University of Washington, Kitsap Lane *ALJ*
Swiss Stone					*Pinus cembra*
4'10"	**43'**	24'	**107**	1995	Kent, 623 W Meeker St *RGB*
Table Mountain					*Pinus pungens*
4'3"	**63'**	28'	**121**	1988	Wind River, USFS Arboretum *RVP*
4'10"	45'	44'	**114**	1993	Lakewood, Tacoma Country Club parking lot *KVP, RVP*
4'9"	33'	**48'**	102	1990	Tacoma, Mountain View Cemetery *KVP, RVP*
Tanyosho					*Pinus densiflora* 'Umbraculifera'
8'4"	32'	**35'**	**141**	1988	Tacoma, Pt Defiance Park *RVP*
5'11"	**33'**	30'	111	1990	Seattle, Kubota Gardens Park *ALJ*
Twisted White					*Pinus strobus* 'Torulosa'
3'7"	**40'**	**19'**	87	1995	Stanwood, 6410 300th St NW *RGB, RVP*
Virginia					*Pinus virginiana*
3'8"	44'	32'	96	1989	Seattle, Washington Park Arboretum, 35N 4W *ALJ, RVP*
Weeping White					*Pinus strobus* 'Pendula'
3'10"	**27'**	24'	79	1992	Puyallup, 404 23rd Ave SE *Martha Robbins*
3'8"	23'	25'	73	1993	Seattle, 2510 29th Ave W *ALJ*

PISTACHIO

Chinese					*Pistacia chinensis*
3'9" largest stem	47'	**43'**	93	1995	Seattle, University Bridge, N end *RGB*

PLANE (see also SYCAMORE, page 93)

Hybrid					*Platanus* x *acerifolia*
20'2"	117'	103'	**385**	1993	Walla Walla, 691 Reser Rd *Shirley Muse*
17'0"	**122'**	95'	350	1993	Walla Walla, 366 S 1st *RVP*
17'3"	106'	**108'**	340	1993	Walla Walla, 1360 Sturn Ave *Shirley Muse*

INTRODUCED TREES

Circumference	Height	Crown Spread	AFA Points	Date Last Measured	Location and Nominators
Oriental					*Platanus orientalis*
11'4"	**92'**	**85'**	249	1993	Seattle, Lake Washington Blvd at E Republican *ALJ*
11'6"	78'	68'	233	1993	Seattle, Salmon Bay Park *ALJ*
Pyramidal Hybrid					*Platanus × acerifolia* 'Pyramidalis'
16'6"	99'	**106'**	323	1993	Walla Walla, Pioneer Park *RVP* (photo below)
15'7"	**105'**	96'	311	1993	Walla Walla, Pioneer Park *RVP*

PLUM (see also **Purple Plums**, page 85)

Circumference	Height	Crown Spread	AFA Points	Date Last Measured	Location and Nominators
Cherry					*Prunus cerasifera*
11'4"	**40'**	**54'**	189	1993	Big Lake, off Hwy 9 on West View Rd *RGB*
Japanese					*Prunus salicina*
6'7"	**34'**	**47'**	125	1995	Seattle, 7311 15th Ave NW *RVP*
7'7"	20'	37'	120	1993	Woodland, 703 S Pekin Rd *RGB, RVP*
Shiro					*Prunus salicina* 'Shiro'
6'3"	**20'**	**39'**	105	1995	Sedro Woolley, 1951 Cook Rd *RGB, RVP*

POPLAR (see also **Hybrid Black Poplars**, page 86)

Circumference	Height	Crown Spread	AFA Points	Date Last Measured	Location and Nominators
Berlin					*Populus × berolinensis*
8'8"	**90'**	**55'**	208	1993	Tacoma, Point Defiance Park *RVP*
Black					*Populus nigra*
7'5"	**71'**	**51'**	175	1993	Seattle, University of Washington, Montlake Cut *ALJ*
Bolleana White					*Populus alba* 'Pyramidalis'
21'7"	110'	67'	386	1992	Vashon Island, 22024 Monument Rd SW *Mike Lee, RGB, RVP*
17'7"	**119'**	57'	344	1992	Vashon Island, 22024 Monument Rd SW *Mike Lee, RGB, RVP*
16'3"	101'	**81'**	316	1987	Seattle, Seattle Pacific University *ALJ*

The **Pyramidal Plane** is common in old European cities and is widely planted in the United States. A hybrid between the American Sycamore and Oriental Plane, it is characterized by horizontal branching and burly trunks. Pioneer Park in Walla Walla has many giants, including our largest (left).

Lombardy Poplars are trees that everyone will recognize. Tall, narrow growth makes them an ideal screen, hence their popularity at drive-in theaters. The tree pictured (right), in Seattle, has outgrown reasonable proportions to become the largest introduced broadleaf tree in the state, as well as the largest known Lombardy Poplar in North America.

Purple Plums

Purple Plums, or Purple-leaved Plums, as they are sometimes more appropriately called, are among the most commonly planted trees in yards and along streets in Washington. Apparently many people like the pleasant pink flowers in spring, followed by rich purple-colored leaves in summer and often rich reds in fall. The fact that many of these leaves turn into a muddy-brown mess for most of the summer and fall, and that many sections of our cities are heavily infested with these dark, somber trees, does little to deter people from planting an ever increasing number. Discovered in Asia Minor around 1880 by a gardener named Pissard, the original Purple Plum is the largest growing, and is most frequently seen in older landscapes. Its small, white flowers are very pleasing in late winter when they appear. Many of the other, more recent cultivars have pink, and/or double flowers which are not subtle but bold and, to some, overpowering. Most Purple Plums find our climate ideal. People interested in learning more about these trees should consult the book *Purpleleaf Plums*, written by Seattle native Arthur Lee Jacobson. Of the tree-sized cultivars, the most common or noteworthy are shown below.

Circumference	Height	Crown Spread	AFA Points	Date Last Measured	Location and Nominators
Burbank Vesuvius					*Prunus* 'Vesuvius'
6'5"	**28'**	**30'**	**112**	1992	Arlington, 59th Ave NE *RGB*
Cistena					*Prunus* 'Cistena'
2'8"	**14'**	**24'**	**52**	1995	Yakima, Yakima Area Arboretum *RGB, ALJ*
Moser					*Prunus* x *blireiana* 'Moseri'
5'0"	26'	31'	**94**	1993	Puyallup, 1829 Pioneer Way *KVP, RVP*
4'7"	**27'**	**33'**	89	1993	Puyallup, 714 2nd Ave NW *KVP, RVP*
Newport					*Prunus* 'Newport'
6'8"	24'	29'	**112**	1993	Puyallup, 1102 15th St NW *KVP, RVP*
6'0"	**25'**	31'	103	1993	Sumner, Sumner Cemetery *RGB, RVP*
5'1"	18'	**35'**	88	1988	Seattle, Evergreen Cemetery *ALJ*
Pissard					*Prunus cerasifera* 'Pissardii'
8'8"	38'	42'	**152**	1993	Sumner, 912 Sumner Ave *KVP, RVP*
7'4"	**48'**	**57'**	150	1987	Seattle, W end of NW 85th St *RVP*
Seattle Hollywood					*Prunus* "Seattle Hollywood"
4'4"	**37'**	**42'**	**99**	1992	Seattle, University of Washington, Herb Garden *ALJ*
Thundercloud					*Prunus cerasifera* 'Thundercloud'
5'4"	30'	**39'**	**104**	1988	Seattle, 16th Ave E & E McGraw *ALJ*
3'5"	**34'**	27'	82	1988	Seattle, Carl S English Gardens, Ballard Locks *ALJ*
Trailblazer					*Prunus* 'Trailblazer'
5'0"	**22'**	**25'**	**88**	1993	Walla Walla, 507 E Maple St *RGB, RVP*

The **Moser Plum**, one of the smaller-growing of the Purple Plums, is nonetheless attractive in bloom. Its parent, the Blireiana Plum, grows too small to include in this book. These, however, are the two common Purple Plums that have double flowers. Our champion Moser Plum is this tree in Puyallup, shown in full bloom.

INTRODUCED TREES

Hybrid Black Poplars

Populus x *canadensis*

Hybrid Black Poplars are a group of crosses developed in Europe during the last century between the American Eastern Cottonwood, *Populus deltoides*, and the European Black Poplar, *Populus nigra*. These fast-growing trees can rapidly reach impressive proportions, and are frequently grown for wood production as well as for ornament. Perhaps the earliest of these hybrids was marketed as the **Carolina Poplar**, and received widespread planting throughout the United States. This particular hybrid is probably the same as 'Eugenei', a hybrid between the Eastern Cottonwood and the Lombardy Poplar, *Populus nigra* 'Italica'. Although some of these hybrids are male and some are female, they are difficult to sort out, requiring close examination during flowering and bud break. There are several types of Hybrid Black Poplars that are commonly encountered in older parks and gardens of Washington, several of the largest of which are listed here.

The trees pictured are **Regenerata Poplars** along Seattle's Mt. Baker Blvd. Regenerata is a female clone, hence drifts of 'snow' which follow flowering in spring. Noted for outstanding fall color, these specimens are worth a visit in October.

Cultivar	Circumference	Height	Crown Spread	AFA Points	Date Last Measured	Location and Nominators
'Eugenei'	**18'6"**	104'	87'	**348**	1988	Wenatchee, 5th & Wenatchee Sts *RVP*
	16'5"	**124'**	**94'**	344	1988	Walla Walla, Pioneer Park *RVP*
'Marilandica'	9'6"	**100'**	**71'**	**232**	1988	Seattle, Washington Park Arboretum, 19N 6W *ALJ*
	10'10"	82'	70'	229	1988	Seattle, Alki Beach *ALJ*
'Regenerata'	12'10"	109'	**84'**	**284**	1988	Seattle, Mt Baker Blvd & 33rd Ave S *ALJ*
	13'0"	**111'**	61'	282	1988	Tacoma, Pt Defiance Park *RVP*
'Robusta'	11'0"	**122'**	**92'**	**277**	1988	Tacoma, Pt Defiance Park *RVP*
	11'10"	85'	66'	243	1988	Seattle, Magnolia Playfield *ALJ*
'Serotina'	10'2"	**102'**	77'	**243**	1993	Tacoma, Point Defiance Park *RVP*
	10'11"	91'	**79'**	242	1993	Seattle, Alki Beach *ALJ*
'Serotina Aurea'	**12'1"**	**99'**	77'	**263**	1992	Puyallup, Linden Golf & Country Club *RGB*
	12'0"	97'	**82'**	261	1992	Puyallup, Linden Golf & Country Club *RGB*
	11'11"	97'	79'	260	1992	Puyallup, Linden Golf & Country Club *RGB*

INTRODUCED TREES

Circumference	Height	Crown Spread	AFA Points	Date Last Measured	Location and Nominators
Cathay					*Populus cathayana*
3'5"	**62'**	**44'**	**114**	1994	Seattle, Washington Park Arboretum, 14N 2E *ALJ*
3'10"	47'	31'	101	1993	Seattle, Camp Long *ALJ*
Certines					*Populus* x *berolinensis* 'Certinensis'
8'5"	**115'**	**54'**	**229**	1988	Seattle, Washington Park Playfield *ALJ*
Columnar Simon					*Populus simonii* 'Fastigiata'
10'4"	**75'**	**51'**	**212**	1993	Royal City, Hwy 26, 4.5 miles W of town *RGB, ALJ*
Ghost					*Populus nigra* 'Afghanica'
10'7"	104'	32'	**239**	1993	Seattle, University of Washington, Montlake Cut *ALJ*
8'0"	**118'**	27'	221	1995	Seattle, Washington Park Arboretum, 32N 7W *KVP, RVP*
Gray					*Populus* x *canescens*
13'6"	**87'**	74'	**267**	1993	Olympia, Masonic Cemetery on Cleveland Ave *RVP*
12'4"	81'	**79'**	249	1992	Steilacoom, Western Washington State Hospital *RVP*
Japanese					*Populus maximowiczii*
5'1"	**72'**	**51'**	**146**	1993	Seattle, Washington Park Arboretum, 15N 0 *RVP*
Lombardy					*Populus nigra* 'Italica'
30'5"	**142'**	**67'**	**524**	1995	Seattle, Lake Washington Bvd & S Atlantic St *ALJ (photo page 84)*
Manchurian					*Populus songarica*
8'1"	**95'**	39'	**202**	1992	Seattle, Washington Park Arboretum, 48N 5E *ALJ*
Szechuan					*Populus szechuanica*
5'8"	**75'**	**53'**	**156**	1993	Seattle, Washington Park Arboretum, 15N 1E *RVP*
6'0"	68'	49'	152	1993	Seattle, Washington Park Arboretum, 15N 1E *RVP*
Weeping Simon					*Populus simonii* 'Pendula'
12'10"	69'	**66'**	**239**	1988	Seattle, Washington Park Arboretum, 49N 7W *ALJ*
8'10"	**101'**	56'	221	1995	Seattle, Washington Park Arboretum, 50N 7W *ALJ*
White					*Populus alba* 'Nivea'
18'0"	**90'**	**92'**	**329**	1988	Allen, 1 mile N on Ershig Rd *RVP*

PRIVET

Chinese					*Ligustrum lucidum*
4'2"	**44'**	24'	**100**	1993	Seattle, 4714 NE 20th *Rich Ellison*
Mock					*Phillyrea latifolia*
2'2"	**24'**	**24'**	**56**	1995	Seattle, Washington Park Arboretum, 11N 9E *RGB*

QUINCE

Chinese					*Pseudocydonia sinensis*
2'2"	**34'**	**28'**	**67**	1990	Seattle, University of Washington, Herb Garden *ALJ*
Common					*Cydonia oblonga*
5'6"	23'	29'	**96**	1990	Seattle, 133 14th Ave E *ALJ*
4'9"	**27'**	**35'**	93	1992	Tacoma, 4530 S K St *RGB, KVP, RVP*
largest stem					

REDBUD

Eastern					*Cercis canadensis*
5'3"	38'	**36'**	**110**	1993	Tacoma, 2901 Soundview Dr *KVP, RVP (photo page 88)*
2'3"	**43'**	29'	77	1993	Seattle, 1900 Shenandoah Dr E *ALJ*
Forest Pansy					*Cercis canadensis* 'Forest Pansy'
2'6"	**19'**	**28'**	**56**	1993	Seattle, 2844 Cascadia Ave S *KVP, RVP*
Judas-tree					*Cercis siliquastrum*
5'8"	**37'**	33'	**113**	1989	Seattle, Broadway Reservoir *ALJ*
5'8"	36'	33'	112	1989	Seattle, 4925 Stanford Ave NE *ALJ*
4'5"	27'	**39'**	90	1993	Seattle, 5525 35th Ave NE *RVP*

INTRODUCED TREES

Circumference	Height	Crown Spread	AFA Points	Date Last Measured	Location and Nominators
REDWOOD					
Blue					*Sequoia sempervirens* **f. glauca**
9'6"	**95'**	48'	**221**	1993	Seattle, Washington Park Arboretum, 39N 7W *RGB*
Cantab					*Sequoia sempervirens* **'Cantab'**
4'5"	**51'**	27'	**111**	1995	Port Angeles, private residence *Jim Causton*
Coast					*Sequoia sempervirens*
19'4"	140'	61'	**392**	1993	Olympia, Daniel J. Evans Park *RVP (photo below)*
14'1"	**149'**	62'	333	1993	Seattle, Interlaken Park *ALJ*
Dawn					*Metasequoia glyptostoboides*
11'7"	67'	41'	**216**	1993	Lakewood, Lakewold Gardens *KVP, RVP (photo page 89)*
9'5"	93'	32'	**214**	1993	Seattle, Washington Park Arboretum, 13N 7E *ALJ*
6'0"	**100'**	19'	177	1992	Tacoma, 3405 N 34th *RGB, KVP, RVP*
REHDER TREE					*Rehderodendron macrocarpum*
2'11"	**38'**	**27'**	**80**	1993	Seattle, Washington Park Arboretum, 12N 7E *ALJ*
RUSSIAN OLIVE					*Elaeagnus angustifolia*
8'3"	50'	51'	**162**	1993	Walla Walla, VA Hospital *Shirley Muse, RGB, RVP*
8'5"	47'	49'	**160**	1995	Richland, Sham-Na-Pum Golf Crs *Mid Columbia Forestry Council*
8'1"	48'	**54'**	**159**	1995	Richland, Sham-Na-Pum Golf Crs *Mid Columbia Forestry Council*
6'8"	**53'**	39'	143	1995	Richland, Sham-Na-Pum Golf Crs *Mid Columbia Forestry Council*
SASSAFRAS					*Sassafras albidum*
8'8"	46'	47'	**162**	1993	Tacoma, 1016 S 37th St *RVP*
8'0"	45'	**49'**	**153**	1995	Tacoma, 1901 S 40th St *KVP, RVP*
2'4"	**74'**	17'	106	1989	Seattle, Seward Park *ALJ*

Kathy is standing with the largest **Eastern Redbud**, in Tacoma. Generally a small tree, redbuds are spectacular at bloom-time in early summer.

The **Coast Redwood** is the world's tallest tree. This giant tree in Olympia is already 140' tall, and had a small city park made to include it.

		Crown	AFA	Date Last	
Circumference	Height	Spread	Points	Measured	Location and Nominators

SEQUOIA

Blue *Sequoiadendron giganteum* 'Glaucum'

16'6"	**119'**	40'	**327**	1992	Tacoma, Point Defiance Park *KVP, RVP*

Giant *Sequoiadendron giganteum*

32'6"	125'	65'	**531**	1995	Ridgefield, 605 N Main *Audrey & Paul Grescoe (photo below)*
17'5"	**157'**	49'	378	1993	Wind River, USFS Arboretum *RVP*

Weeping *Sequoiadendron giganteum* 'Pendulum'

7'10"	40'	**35'**	**143**	1995	Seattle, 4518 SW Wildwood Dr *ALJ*
2'9"	**44'**	12'	80	1992	Seattle, Evergreen Cemetery *RVP*

SERVICEBERRY

Allegheny *Amelanchier laevis*

1'5"	**42'**	20'	**64**	1995	Seattle, Washington Park Arboretum, 26N 0 *RGB*

SERVICE TREE (see also **MOUNTAIN ASH**, page 74; and **WHITEBEAM**, page 97)

Devon *Sorbus devoniensis*

4'0"	**62'**	**37'**	**119**	1995	Seattle, Washington Park Arboretum, 22N 4E *KVP, RVP*

Service Tree of Fontainbleau *Sorbus* x *latifolia*

3'10"	**53'**	**33'**	**107**	1993	Seattle, Washington Park Arboretum, 20N 4E *ALJ*

True *Sorbus domestica*

9'5"	77'	**61'**	**205**	1993	Edmonds, 11303 S Dogwood Lane *RGB*
4'5"	**78'**	37'	140	1993	Seattle, Washington Park Arboretum, 36N 4E *ALJ*

Wild *Sorbus torminalis*

3'5"	**42'**	**35'**	**92**	1995	Seattle, Washington Park Arboretum, 21N 5E *KVP, RVP*

The **Dawn Redwood** is a Chinese tree thought extinct until rediscovered in 1941. Like its cousins the Coast Redwood and Giant Sequoia, this tree has great vigor. The tree pictured is at Lakewold Gardens in Lakewood, and is already 3½ feet in diameter.

The **Giant Sequoia**, the world's largest tree, needs no introduction. This species likes our climate; it is our tallest introduced tree, the largest, and has the largest diameter. Our largest Sequoia, pictured above, is a newly discovered tree in Ridgefield.

INTRODUCED TREES

	Circumference	Height	Crown Spread	AFA Points	Date Last Measured	Location and Nominators

SILK TREE
Albizia julibrissin

	7'2"	46'	**75'**	**151**	1990	Vancouver, Community Library on Ft Vancouver Way *RVP*
	4'5"	**49'**	47'	114	1993	Woodland, Hulda Klager Lilac Garden *RGB, RVP*

SILVERBELL

Mountain
Halesia carolina **var. monticola**

	6'6"	63'	47'	**153**	1990	Tacoma, 420 N Sheridan *KVP, RVP*
	3'11"	**68'**	**53'**	128	1992	Lakewood, Lakewold Gardens *RGB, RVP*

SMOKE TREE

American
Cotinus obovatus

	4'1"	**49'**	**31'**	**106**	1992	Lakewood, Lakewold Gardens *KVP, RVP*

largest of 2 trunks

European
Cotinus coggygria

	3'7"	**35'**	**35'**	87	1993	Walla Walla, 3rd and Tietan St *Shirley Muse, RGB, RVP*

largest of 4 stems

Purple
Cotinus coggygria **'Foliis Purpureis'**

	3'0"	**23'**	**28'**	66	1993	Seattle, 2844 Cascadia Ave S *KVP, RVP*

SNOWBELL

Bigleaf
Styrax obassia

	2'8"	48'	**25'**	**86**	1990	Seattle, Washington Park Arboretum, 35N 0 *ALJ, RVP*
	3'5"	30'	21'	76	1990	Seattle, Washington Park Arboretum, 18N 2W *ALJ*
	1'10"	**50'**	22'	77	1990	Seattle, 10723 24th Ave NE *RGB*

Japanese
Styrax japonicus

	4'4"	**40'**	30'	**99**	1988	Seattle, Broadway E, N of E Lynn St *ALJ*
	4'6"	29'	32'	91	1988	Seattle, 1119 37th Ave E *ALJ*
	2'9"	49'	**33'**	90	1990	Seattle, 10723 24th Ave NE *RGB*

SORRELTREE
Oxydendrum arboreum

	4'1"	38'	**25'**	**93**	1990	Tacoma, Jefferson Park *RVP*
	1'11"	**48'**	14'	74	1989	Seattle, Washington Park Arboretum, 12N 8E *ALJ*

SOUTHERN BEECH

Antarctic
Nothofagus antarctica

	4'4"	**55'**	**33'**	**115**	1990	Seattle, Washington Park Arboretum, 49N 3E *ALJ*

Dombey
Nothofagus dombeyi

	3'8"	**86'**	35'	**139**	1988	Seattle, Washington Park Arboretum, 42N 1W *ALJ*
	4'1"	71'	**36'**	129	1988	Seattle, Washington Park Arboretum, 11N 4E *ALJ*

Roble
Nothofagus obliqua

	4'8"	67'	**38'**	**132**	1990	Seattle, Washington Park Arboretum, 49N 3E *ALJ*

SPINDLE TREE

European
Euonymus europaeus

	4'0"	18'	**31'**	**74**	1992	Seattle, Roanoke Park *ALJ*

below forking

	2'3"	**25'**	17'	56	1992	Woodinville, Chateau Ste Michelle Winery *RGB, RVP*

Japanese
Euonymus hamiltonianus **ssp. sieboldianus**

	3'10"	18'	**32'**	72	1991	Kirkland, 12010 120th Pl NE, S of Motel 6 *RGB*

Circumference	Height	Crown Spread	AFA Points	Date Last Measured	Location and Nominators
SPRUCE					
Black					*Picea mariana*
5'3"	49'	25'	**118**	1995	Brinnon, Whitney Gardens *RGB, RVP*
2'5"	**51'**	17'	84	1995	Wind River, USFS Arboretum *RVP*
Blue					*Picea pungens* **f.** *glauca*
7'6"	96'	31'	**194**	1988	Spokane, 2105 Rockwood Blvd *ALJ, RVP (photo below)*
5'6"	**102'**	25'	174	1988	Spokane, Rockwood Blvd & Garfield *ALJ, RVP*
Brewer					*Picea breweriana*
3'6"	**52'**	23'	**100**	1990	Wind River, USFS Arboretum *RVP*
Columnar Blue					*Picea pungens* 'Columnaris'
2'6"	32'	4'	**63**	1993	Tacoma, 110 S 68th St *KVP, RVP*
Cypress Norway					*Picea abies* 'Cupressina'
2'7"	**46'**	19'	**82**	1995	Seattle, Washington Park Arboretum, 26N 4E *RGB*
Dragon					*Picea asperata*
4'10"	**70'**	33'	**136**	1995	Seattle, Washington Park Arboretum, 26N 4E *RVP*
Golden Norway					*Picea abies* 'Aurea'
4'3"	**39'**	33'	**101**	1995	Mt Vernon, 1311 McLean Rd *RGB*
Hondo					*Picea jezoensis* **var.** *hondoensis*
4'8"	44'	40'	**110**	1988	Seattle, Evergreen Cemetery *ALJ*
3'8"	**50'**	31'	102	1988	Bremerton, Evergreen Park *ALJ, RVP*
Norway					*Picea abies*
11'8"	91'	43'	**242**	1995	Bellingham, Carl Cozier Elem School *RGB, RVP*
10'2"	101'	59'	**238**	1993	Olympia, across from the Capitol Museum *RVP*
10'4"	103'	45'	**238**	1995	Tacoma, Pt Defiance Park *RVP*
8'2"	**118'**	45'	227	1988	Spokane, Finch Arboretum *ALJ, RVP*

Although not native to Washington, the **Blue Spruce** finds itself quite at home in the environment of eastern Washington. All of our largest, including this one, may be found in Spokane.

Of all the Spruce species that grow in Washington, one of the loveliest is the **Oriental Spruce.** Native to the mountains of western Asia, the tiny needles and rich green color of this spruce make it easy to identify.

*INTRODUCED TREES*_____

Circumference	Height	Crown Spread	AFA Points	Date Last Measured	Location and Nominators
Oriental					*Picea orientalis*
9'7"	85'	39'	**210**	1993	Seattle, Lakeview Park *ALJ*
9'1"	90'	40'	**209**	1993	Puyallup, 5811 124th Ave Ct E *RGB, RVP*
8'6"	**91'**	39'	**203**	1993	Tacoma, Pt Defiance Park *RVP (photo page 91)*
Pyramidal Norway					**(?)** *Picea abies* **f. pyramidata**
7'10"	60'	39'	**164**	1993	Tacoma, 3324 N 30th *RVP*
Red					*Picea rubens*
3'10"	51'	25'	**103**	1995	Clearview, 17927 Hwy 9 *RGB, RVP*
3'4"	**51'**	29'	**98**	1990	Wind River, USFS Arboretum *RVP*
Sakhalin					*Picea glehnii*
4'0"	58'	35'	**115**	1995	Seattle, Washington Park Arboretum, 25N 5E *RGB, RVP*
Sargent					*Picea brachytyla*
3'3"	59'	19'	**103**	1995	Seattle, Washington Park Arboretum, 25N 5E *RGB, RVP*
Serbian					*Picea omorika*
5'4"	61'	21'	**130**	1992	Lake Stevens, 5224 95th Ave NE *RGB*
Snakebranch					*Picea abies* **f. virgata**
5'0"	73'	32'	**141**	1992	Seattle, Volunteer Park *ALJ (photo below)*
Tapo Shan					*Picea asperata* **var. retroflexa**
3'2"	60'	23'	**104**	1995	Seattle, Washington Park Arboretum, 25N 5E *RGB, RVP*
Tigertail					*Picea polita*
2'4"	46'	13'	**77**	1993	Seattle, Washington Park Arboretum, 33N 6W *RVP*
Twisted Branch Norway					*Picea abies* **'Intermedia'**
4'5"	65'	20'	**123**	1995	Bainbridge Island, 10451 Arrow Pt Dr NE *RGB*

Pictured here are two of our bizarre forms of Norway Spruce: the **Snakebranch Spruce** (left), and the **Weeping Norway Spruce** (right). These illustrate the diversity of tree types that are grown in Washington. Not only do we have a high diversity, but the growth rates are high. Very few of our exotic trees are older than 100 years, yet many have outgrown older trees in the eastern United States or Europe.

Circumference	Height	Crown Spread	AFA Points	Date Last Measured	Location and Nominators
Weeping Norway					*Picea abies* 'Inversa'
4'3"	**52'**	**27'**	**110**	1992	Puyallup, Woodbine Cemetery *KVP, RVP (photo page 92)*
West Himalayan					*Picea smithiana*
2'10"	**42'**	20'	**81**	1995	Seattle, University of Washington, Friendship Grove *RGB*
3'0"	38'	24'	**80**	1995	Seattle, Washington Park Arboretum, 26N 5E *RGB, RVP*
White					*Picea glauca*
3'11"	63'	28'	**117**	1989	Seattle, Washington Park Arboretum, 35N 4W *ALJ, RVP*
3'4"	**65'**	18'	109	1987	Seattle, Washington Park Arboretum, 30N 6W *ALJ*
4'6"	34'	33'	96	1992	Tacoma, University of Puget Sound *KVP, RVP*
Yeddo					*Picea jezoensis*
5'2"	**49'**	39'	121	1990	Seattle, 9733 Arrowsmith Ave S *ALJ*

STEWARTIA

Circumference	Height	Crown Spread	AFA Points	Date Last Measured	Location and Nominators
Common					*Stewartia pseudocamellia*
3'5"	**45'**	29'	**93**	1992	Puyallup, 15014 106th St E *RGB, RVP*
2'6"	**45'**	**33'**	83	1989	Seattle, Washington Park Arboretum, 10N 5E *ALJ*
Mountain					*Stewartia ovata*
1'8"	**26'**	**17'**	❖50	1995	Seattle, Washington Park Arboretum, 13N 9W *ALJ*
Tall					*Stewartia monadelpha*
2'11"	**41'**	**35'**	85	1990	Seattle, Washington Park Arboretum, 10N 5E *ALJ*

SUMAC

Circumference	Height	Crown Spread	AFA Points	Date Last Measured	Location and Nominators
Pottanin					*Rhus pottaninii*
3'2"	**35'**	**31'**	81	1995	Seattle, Cowen Park *RGB*
Staghorn					*Rhus typhina*
3'2"	31'	**25'**	75	1990	Seattle, Univ of Wash, Oceanography Library *ALJ, RVP*
2'7"	**41'**	12'	75	1990	Seattle, Univ of Wash, Oceanography Library *ALJ, RVP*

SWEETGUM

Circumference	Height	Crown Spread	AFA Points	Date Last Measured	Location and Nominators
American					*Liquidambar styraciflua*
8'10"	100'	65'	**222**	1988	Walla Walla, Pioneer Park *RVP*
7'9"	**109'**	50'	**214**	1993	Walla Walla, Pioneer Park *RVP*
9'9"	83'	53'	**213**	1993	Olympia, Capitol Museum *RVP*
9'0"	83'	75'	**213**	1993	Seattle, Broadmoor, N entrance *Ian McCallum*
6'7"	69'	**79'**	168	1990	Tacoma, Pt Defiance Park *RVP*
Golden					*Liquidambar styraciflua* 'Variegata'
5'5"	**66'**	**47'**	143	1995	Seattle, 419 31st Ave S *RGB*
Oriental					*Liquidambar orientalis*
4'8"	**30'**	**27'**	93	1993	Seattle, Washington Park Arboretum, 12N 3W *RVP*

SYCAMORE (see also **PLANE**, page 83)

Circumference	Height	Crown Spread	AFA Points	Date Last Measured	Location and Nominators
American					*Platanus occidentalis*
13'5"	95'	**99'**	**281**	1990	Walla Walla, E Pleasant St & Berney Dr *Shirley Muse*
10'6"	**112'**	75'	257	1988	Walla Walla, Pioneer Park *RVP*

TAMARISK

Circumference	Height	Crown Spread	AFA Points	Date Last Measured	Location and Nominators
Spring-Flowering					*Tamarix parviflora*
4'7"	**25'**	**21'**	85	1995	Woodland, 1123 S Pekin Rd *RGB, RVP*
Summer-Flowering					*Tamarix chinensis* 'Plumosa'
5'7"	**26'**	**29'**	100	1995	Seattle, 634 11th Ave E *RGB*

largest of 4 trunks

INTRODUCED TREES

Circumference	Height	Crown Spread	AFA Points	Date Last Measured	Location and Nominators
TANOAK					*Lithocarpus densiflorus*
3'11"	**60'**	**50'**	**119**	1993	Richmond Beach, 20066 15th Ave NW *Art Kruckeberg*
largest of 2 stems					
4'9"	53'	33'	**118**	1993	Richmond Beach, 20066 15th Ave NW *Art Kruckeberg*
Cutleaf					*Lithocarpus densiflorus* **f. attenuato-dentatus**
2'10"	**40'**	**25'**	**80**	1993	Richmond Beach, 20066 15th Ave NW *Mareen Kruckeberg*
TORREYA					
Japanese					*Torreya nucifera*
3'6"	**34'**	**33'**	**84**	1992	Seattle, 4233 Ashworth Ave N *ALJ*
largest of 3 trunks; topped					
TREE OF HEAVEN					*Ailanthus altissima*
13'10"	75'	51'	**254**	1990	Vancouver, 2303 Kauffman Ave *RVP*
11'3"	85'	**79'**	**240**	1988	Walla Walla, Whitman College *RVP (photo below)*
6'11"	**89'**	51'	185	1988	Walla Walla, Pioneer Park *RVP*
TULIPTREE					*Liriodendron tulipifera*
19'7"	94'	**79'**	**349**	1993	Mount Vernon, 1003 Cleveland St *RVP*
13'2"	**121'**	**79'**	297	1990	Seattle, 700 McGilvra Blvd E *ALJ (photo below)*
8'6"	**121'**	47'	234	1993	Tacoma, Wright Park *RVP*

A native of China, the **Tree of Heaven** is one of the most prolific of our introduced trees. Now 'native' in almost every state, it has smooth bark and attractive seed clusters. The tree pictured is at the Whitman College campus in Walla Walla.

Although not our largest, at 121 feet this **Tuliptree** in Seattle is our tallest. Among the eastern United States' largest and tallest tree species, Tuliptrees have unusual leaves that turn a spectacular gold in autumn.

Circumference	Height	Crown Spread	AFA Points	Date Last Measured	Location and Nominators

TUPELO

Black
Nyssa sylvatica

5'1"	**57'**	**45'**	**129**	1990	Tacoma, Jefferson Park *KVP, RVP*
5'4"	54'	39'	**128**	1987	Seattle, 14th Ave E, N of E Roy St *ALJ*

UMBRELLA-PINE

Japanese
Sciadopitys verticillata

10'7" below forking	45'	**31'**	**180**	1990	Woodland, Hulda Klager Lilac Garden *RGB*
7'5"	**54'**	27'	150	1987	Everett, Evergreen Cemetery *ALJ, RVP*

WALNUT (see also **BUTTERNUT**, page 33)

Arizona
Juglans major

4'9"	**65'**	**51'**	**135**	1990	Seattle, 1911 14th Ave E *ALJ*

Black
Juglans nigra

19'2"	95'	**119'**	**354**	1995	Skagit City, 1942 Skagit City Rd *RVP (photo page 100)*
8'8"	**112'**	78'	235	1988	Walla Walla, Pioneer Park *RVP*

Cathay
Juglans cathayensis

17'8" below forking	**76'**	83'	**309**	1993	Issaquah, 885 2nd Ave SE *RGB*
10'6"	69'	**105'**	221	1993	Auburn, 1201 4th St NE *RGB, RVP*

The tree pictured above is one of our most perfectly formed trees. It is an **English Walnut**, growing out in a field in the Wynoochee River Valley. Generally much smaller than the Black Walnuts, a few of our English Walnuts have grown into impressive trees. English Walnuts produce the nuts that most of us are familiar with. With nine or so leaflets, this species has fewer leaflets than any other walnut.

INTRODUCED TREES

Circumference	Height	Crown Spread	AFA Points	Date Last Measured	Location and Nominators
English					*Juglans regia*
12'5"	76'	83'	**246**	1992	Fall City, W Snoqualmie River Rd & SE 19th Way *RGB*
12'4"	68'	72'	**234**	1992	Carnation, 31803 Eugene St *RGB*
10'8"	**82'**	63'	226	1992	Arlington, ½ mile E of I-5 on State Route 530 *RGB, RVP*
11'7"	66'	**91'**	228	1993	Montesano, 1168 Wynoochee Valley Rd *RVP (photo page 95)*
Heartnut					*Juglans ailanthifolia* **var.** *cordiformis*
6'4"	**37'**	**64'**	**129**	1992	Puyallup, 5728 River Rd, at 58th Ave E *KVP, RVP*
Hybrid English					*Juglans* × *intermedia*
13'2"	**84'**	**81'**	**262**	1990	Vancouver, 6106 SE Riverside Dr *RVP*
Japanese					*Juglans ailanthifolia*
15'9" below forking	49'	72'	**256**	1992	Ferndale, 7857 Enterprise Rd *RGB*
9'6"	**57'**	77'	190	1993	Olympia, 316 20th Ave *RVP*
10'7"	55'	**79'**	202	1993	Snohomish, 516 Avenue B, Sprauge House, in alley *RVP*
Manchurian					*Juglans mandshurica*
8'1"	**45'**	62'	**157**	1992	Puyallup, 815 4th St SW *RGB, RVP*
8'0"	42'	**63'**	154	1992	Puyallup, 1402 21st St SE *RGB*

Planted by F.S. Martin in 1908, this walnut in Puyallup is a very unusual type. The tree was given to Mr. Martin by his good friend Luther Burbank. Burbank became famous early in this century by developing many new types of trees through hybridization. The tree pictured was one he developed for wood production, and is a hybrid between the eastern Black Walnut and the English Walnut. He named it the **Paradox Walnut** because it was common to hybridize walnuts to improve one or more characteristics of nut production, but this tree puts very little energy into its nuts, which are small and nearly meatless. This tree is on property recently purchased by the high school. I hope the historical significance of this noble tree will be recognized and protected.

Circumference	Height	Crown Spread	AFA Points	Date Last Measured	Location and Nominators
Northern California Black					*Juglans hindsii*
15'4"	**102'**	**105'**	**312**	1993	Vancouver, mile marker 6 on Lower River Rd *RVP*
16'6"	79'	80'	**297**	1988	Yakima, S 28th Ave at Palatine Ave *RVP*
Paradox					*Juglans* 'Paradox'
12'9"	73'	**75'**	**245**	1993	Puyallup, 517 7th St SW *RGB, RVP (photo page 96)*
Royal Hybrid					*Juglans hindsii* x *nigra*
13'0"	**88'**	79'	**264**	1990	Tacoma, Tacoma Ave at S 4th St *KVP, RVP*
13'1"	84'	**85'**	262	1988	Seattle, 22nd Ave & E James St *ALJ*
12'3"	**88'**	75'	254	1995	Port Townsend, Chief Chetzemoka Park *RVP*
Texas Black					*Juglans microcarpa*
6'9"	45'	**64'**	**142**	1988	Seattle, Washington Park Arboretum, 54N 3E *ALJ*

WHITEBEAM (see also **MOUNTAIN ASH**, page 74; and **SERVICE TREE**, page 89)   *Sorbus aria*

Circumference	Height	Crown Spread	AFA Points	Date Last Measured	Location and Nominators
3'3"	**42'**	**29'**	**88**	1993	Seattle, Washington Park Arboretum, 20N 4E *ALJ*
Finnish					*Sorbus* x *thuringiaca*
9'8"	**56'**	**53'**	**185**	1993	Brush Prairie, 16017 NE Caples *RGB*
Swedish					*Sorbus intermedia*
9'10"	26'	36'	**153**	1990	Tacoma, 1104 N 27th *RVP (photo below)*
3'11"	**52'**	35'	108	1992	Parkland, Pacific Lutheran University, SE corner Stuen Hall *RVP*
largest of 2 stems					
5'5"	46'	**39'**	121	1988	Tacoma, Wright Park *RVP*
Wilfrid Fox					*Sorbus* 'Wilfrid Fox'
3'2"	**47'**	**29'**	**92**	1993	Seattle, Washington Park Arboretum, 20N 4E *ALJ*

While the European Mountain Ash is a common feature of our urban landscapes, many of its relatives are not. This is a pity, as the *Sorbus* group, which comprises the Mountain-Ashes, Service Trees, and Whitebeams, contains many attractive trees. In addition, our climate is ideally suited for this group. The *Sorbus* collection at the Washington Park Arboretum in Seattle is world famous. The tree pictured is our Champion **Swedish Whitebeam** in Tacoma. While topped for the power line, it is nonetheless a glorious sight in full bloom.

INTRODUCED TREES

		Crown	AFA	Date Last	
Circumference	Height	Spread	Points	Measured	Location and Nominators

WILLOW

Bay
Salix pentandra

6'10"	37'	43'	130	1990	Seattle, 55204 NE 80th St *ALJ*

Chinese
Salix matsudana

7'11"	44'	61'	154	1992	Seattle, Madrona Park *ALJ*

Chinese Weeping
Salix babylonica

8'10"	33'	53'	152	1995	Seattle, University of Washington, Fish Rearing Pond *RVP*
9'1"	30'	45'	150	1995	Seattle, University of Washington, Fish Rearing Pond *RVP*

Corkscrew
Salix matsudana 'Tortuosa'

12'10"	56'	71'	228	1993	Grandview, 606 E 3rd St *RGB, RVP*

Golden
Salix alba var. *vitellina*

19'3"	77'	84'	329	1988	Spokane, Manito Park *ALJ, RVP*
largest of 2 stems					
16'2"	83'	72'	295	1992	Stanwood, Marine Dr near Florence Rd *RGB*

Golden Weeping
Salix x *sepulcralis* 'Chrysocoma'

16'8"	63'	57'	277	1992	Fife, end of 3500 blk 12th St E *KVP, RVP*
14'10"	70'	87'	270	1995	Kennewick, 700 S Monroe St *Mid Columbia Forestry Council*
16'9"	42'	73'	261	1992	Fife, end of 3500 blk 12th St E *KVP, RVP*
13'9"	58'	91'	246	1993	Kent, 7011 S 182nd St *RVP*

Gray
Salix cinerea

11'10"	36'	42'	188	1995	Elma, 1186 Monte Elma Rd *RGB*

Hybrid Pussy
Salix x *sericans*

3'8"	33'	33'	85	1995	Redmond, 13245 Woodinville-Redmond Rd NE *Mike Lee*

Hybrid Weeping
Salix x *sepulcralis* 'Salamonii'

14'11"	83'	75'	281	1990	Fife, 500' S of tracks on 54th Ave S *KVP, RVP*
12'0"	85'	87'	251	1988	Seattle, Madrona Park *ALJ*

Hybrid White
Salix x *rubens*

12'8"	89'	55'	255	1993	Seattle, Washington Park Arboretum, 47N 4E *ALJ*
13'2"	61'	49'	231	1993	Seattle, Green Lake Park *ALJ*

Narrow-Leaved Rosemary
Salix elaeagnos ssp. *angustifolia*

7'7"	19'	32'	118	1995	Seattle, University of Washington, Fish Rearing Pond *RGB*

Ringleaf
Salix babylonica 'Annularis'

7'10"	52'	61'	161	1990	Tacoma, Pt Defiance Park *KVP, RVP* (photo page 99)

Silver
Salix alba var. *sericea*

10'4"	96'	85'	241	1995	Seattle, Washington Park Arboretum, 47N 7E *ALJ, RVP*

Violet
Salix daphnoides

7'2"	45'	53'	144	1995	Seattle, Washington Park Arboretum, 47N 4E *RGB*

White
Salix alba var. *coerulea*

11'10"	74'	65'	232	1990	Seattle, Washington Park Arboretum, 52N 3E *ALJ*

WINGNUT

Caucasian
Pterocarya fraxinifolia

5'1"	54'	43'	126	1992	Seattle, Washington Park Arboretum, 31N 4W *RVP*

Chinese
Pterocarya stenoptera

6'0"	66'	67'	155	1990	Seattle, Washington Park Arboretum, 29N 3W *RVP*

YELLOWWOOD
Cladrastis kentukea

7'8"	61'	60'	168	1988	Spokane, Coeur d'Alene Park *ALJ, RVP*
7'8"	56'	53'	161	1992	Puyallup, 7613 Stewart Ave E *RGB, RVP*
4'10"	71'	41'	139	1990	Seattle, 10723 24th Ave NE *RGB*

Japanese
Cladrastis platycarpa

3'9"	50'	49'	107	1995	Seattle, Washington Park Arboretum, 18N 7E *RVP*

Circumference	Height	Crown Spread	AFA Points	Date Last Measured	Location and Nominators
YEW					
English					*Taxus baccata*
13'10" below forking	47'	59'	**226**	1992	Puyallup, 1007 13th St SE *KVP, RVP*
6'6"	**55'**	45'	144	1989	Seattle, Cowen Park *ALJ*
7'6"	48'	**60'**	154	1992	Tukwila, 13243 40th Ave S *RGB*
Golden English					*Taxus baccata* 'Aurea'
5'7"	**27'**	**39'**	**104**	1995	Olympia, 912 Olympia Ave *RGB*
Golden Irish					*Taxus baccata* 'Fastigiata Aurea'
–	**28'**	19'	–	1993	Fir Island, 1426 Polson Rd *RGB*
Hybrid					*Taxus* x *media*
3'10" largest stem	40'	36'	95	1995	Seattle, Acacia Cemetery *ALJ*
Irish					*Taxus baccata* 'Fastigiata'
–	**37'**	15'	–	1988	Seattle, 600 35th Ave *RVP*
Shortleaf					*Taxus baccata* 'Adpressa'
3'6" one of many stems	42'	**48'**	96	1995	Tacoma, Wright Park *KVP, RVP*
Spreading English					*Taxus baccata* 'Repandans'
5'1" one of 7 stems	25'	32'	**94**	1995	Bellingham, Broadway Park *RGB*
2'8" one of 4 stems	17'	**37'**	58	1995	Anacortes, Causland Park *RGB, RVP*
ZELKOVA					*Zelkova serrata*
8'5"	**56'**	54'	**170**	1992	Enumclaw, 28216 SE 432nd St *RGB*

Many of our common ornamental willows have some blood from the Chinese Weeping Willow (*Salix babylonica*), although the true Chinese Weeping Willow itself is quite rare in Washington. Most of our common Weeping Willows are either the Hybrid or Golden Weeping varieties. The **Ringleaf Willow,** with its odd curled-up leaves, is a cultivar of the Chinese Weeping Willow. Point Defiance Park in Tacoma is where you'll find our largest.

Katy and Jeannie are dwarfed by their giant **Black Walnut**. This tree is located near the Skagit River Delta and is Washington's largest. Although native to the eastern United States, Black Walnuts must like our climate, since the largest in the country are all located in California, Oregon, and Washington.

QUICK REFERENCE

WASHINGTON'S TALLEST INTRODUCED TREES

Giant Sequoia	157'
Coast Redwood	149'
Lombardy Poplar	142'
Red Oak	132'
Silver Maple	127'
Eugenei Poplar	124'
Black Walnut (cut down)	123'
American Elm	122'
Hybrid Plane	122'
Robusta Poplar	122'
European Beech	121'
Tuliptree	121'
Pecan	120'
Copper Beech	119'
Bolleana Poplar	119'
Blue Sequoia	119'
Wych Elm	118'
Norway Spruce	118'
Ghost Poplar	118'
Scarlet Oak	117'
Jeffrey Pine	117'
Blue Atlas Cedar	116'
Greek (Apollo) Fir	116'
Certines Poplar	115'

Scarlet Oaks have not only the noble architecture that helped place the oak so prominently in human history, but also spectacular fall colors. Visit Washington Way in Longview for the best collection in the state, including this tree, our co-champion.

This **Western Redcedar** stump is located at an I-5 rest area near Arlington, and is probably our largest stump. It was relocated from its original position a few miles away on the floodplain of the Stillaguamish River. At twenty feet in diameter, this stump reminds us of the forest conditions that confronted the first settlers of Washington more than a century ago. Although we do have a few Redcedars remaining that rival or exceed this tree in size (see page 3), the roof and hollowed center give this stump a scale that instills in us a sense of this species' immortality.

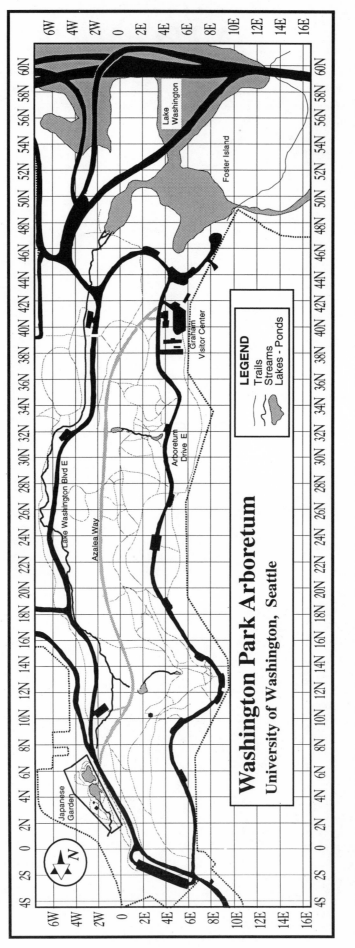

Washington Park Arboretum
University of Washington, Seattle

LEGEND
Trails
Streams
Lakes - Ponds

Japanese Garden

Lake Washington Blvd E

Azalea Way

Arboretum Drive E

Graham Visitor Center

Lake Washington

Foster Island

The Washington Park Arboretum is the site of many trees listed in the Introduced Trees section of this book. The area was originally a city park and nursery for trees to be planted elsewhere in the city. Between 1900 and 1920, the bulk of the park was turned over to the University for use as an arboretum. The Olmsted firm of landscape architects designed Lake Washington Blvd. which passes through the Arboretum, and many of the plantings from 1936 through the 1940's followed their designs. The Arboretum now contains nearly 5,000 taxa in a 200 acre park, and is the destination of many people, either to study tree identification, or just to enjoy the beauty at any time of year.

*FORMER RECORD TREES*_____

Trees come and trees go, but luckily we find more than we lose. Below is a listing of trees lost since 1988. Actually, many more trees have been lost than are listed here. Most would not be in the current edition were they still alive, due to the discovery of new, larger trees. The trees listed below would be in this edition had they not died or been cut down. Figures in **bold** represent noteworthy dimensions.

	Circumference	Height	Crown Spread	AFA Points	Location
NATIVE TREES					
Bitter Cherry	**9'2"**	**104'**	**45'**	**225**	Vashon Island
This tree pre-dates my participation, and could not be located.					
Pacific Dogwood	10'9"	**80'**	55'	**223**	Seattle, Discovery Park
Tree fell over, summer 1994.					
Western Dogwood	3'2"	**23'**	**21'**	66	Woodland
Tree fell over, December 1995.					
Alpine Larch	**22'1"**	**101'**	**84'**	**387**	Wenatchee NF, Carne Mtn trail
Tree fell over, summer 1992.					
INTRODUCED TREES					
Oriental Arborvitae	**5'7"**	38'	23'	**111**	Seattle
Tree cut down by owner, unknown reason.					
Green Ash	9'0"	**98'**	75'	225	Seattle, Japanese Garden
Tree cut down because of shade.					
Box Elder	11'6"	**71'**	88'	**231**	Olympia, Sylvester Park
Tree cut down by owner, unknown reason.					
Incense Cedar	8'2"	**110'**	28'	215	Tacoma, Pt Defiance Park
Tree cut down, unknown reason.					
Bird Cherry	5'4"	44'	41'	**118**	Seattle
Tree cut down by owner, unknown reason.					
Oshima Cherry	4'4"	**45'**	39'	107	Seattle, Washington Park Arboretum
Tree cut down, November 1994.					
Taoyoma Cherry	3'0"	**24'**	39'	70	Seattle, Washington Park Arboretum
Tree cut down, November 1994.					
Ukon Cherry	**7'5"**	29'	53'	**131**	Seattle, University of Washington
Tree cut down, winter 1996, unknown reason.					
Triomf van Boskoop Cypress	8'2"	**102'**	20'	**205**	Olympia, Capitol Museum
Row of tall trees was cut down at the Capitol Museum.					
Monterey Cypress	17'11"	**103'**	64'	334	Gig Harbor
Awesome tree, removed by homeowner, unknown reason.					
Veitch Fir	**5'4"**	50'	24'	120	Seattle, Green Lake Park
Tree died, winter 1994.					
English Elm	**16'3"**	**114'**	**88'**	**331**	Puyallup
One of largest elms, died of Dutch Elm Disease, summer 1995					
Hybrid Elm	**12'6"**	**92'**	**72'**	**260**	Seattle, Hiawatha Playfield
Interesting tree, unique, blew over					
Cider Gum Eucalyptus	**3'8"**	**62'**	**23'**	**112**	Seattle, University of Washington
Froze, winter 1992.					
Spinning Gum Eucalyptus	**8'0"**	44'	**49'**	**152**	Seattle
	5'1"	**47'**	40'	118	Seattle
Both trees froze, winter 1992.					
Thornless Honeylocust	6'2"	84'	73'	**176**	Seattle, Volunteer Park
Fell, Inauguration Day storm, 1993.					
Bigleaf Linden	**14'4"**	70'	59'	**257**	Skagit City
Tree blew over, December 1995.					

FORMER RECORD TREES

	Circumference	Height	Crown Spread	AFA Points	Location
Pink Locust	8'0"	**74'**	57'	**184**	Seattle, Volunteer Park
Fell, Inauguration Day storm, 1993.					
Evergreen Magnolia	5'1"	**47'**	35'	117	Seattle
Cut down by homeowner.					
Silver Maple	22'3"	89'	112'	**384**	Yakima
Tree split and was cut down.					
Black Oak	13'10"	105'	85'	292	Arlington
Giant tree, cut down for road widening.					
Pin Oak	11'6"	**115'**	75'	272	Seattle, Mt Baker Park
Lofty tree, it developed a split and was cut.					
Austrian Pine	12'3"	76'	**61'**	238	Walla Walla, Whitman College
Building was built 8' from trunk; tree promptly died.					
Coulter Pine	13'10"	57'	57'	237	Seattle
Gorgeous tree, cut down after a branch broke off.					
Jack Pine	3'9"	58'	19'	**108**	Seattle, Volunteer Park
Tree died, 1995.					
Knobcone Pine	10'4"	67'	53'	204	Tacoma
Cut down by homeowner.					
Limber Pine	10'7"	60'	**48'**	199	Des Moines
Gorgeous tree, cut down by homeowner.					
Monterey Pine	11'11"	stump only			Wauna
Tree was cut days before I saw it, was much larger than any others seen in Washington.					
Sugar Pine	9'1"	78'	33'	195	Seattle, University Public Library
Killed by blister rust in 1988.					
Eugenei Poplar	18'7"	113'	83'	357	Blaine
Giant tree, cut down by homeowner.					
Serotina Poplar	10'7"	92'	72'	237	Seattle, Alki Beach Park
Fell, Inauguration Day storm, 1993.					
Wild Service Tree	4'8"	**53'**	47'	121	Seattle, Washington Park Arboretum
Tree fell over.					
Blue Spruce	8'5"	77'	39'	188	Puyallup
Nice tree, cut down by homeowner.					
Staghorn Sumac	3'11"	31'	33'	86	Seattle, City Light
Died of unknown causes.					
Tanoak	5'10"	44'	32'	122	Tacoma
Fantastic specimen, lot subdivided, tree removed.					
Black Walnut	14'0"	**123'**	72'	309	Vancouver
One of our tallest introduced trees, lovely, cut down, unknown reason.					
Bay Willow	10'3"	**43'**	63'	182	Seattle
Tree cut down, unknown reason.					
Chinese Weeping Willow	11'5"	52'	64'	207	Steilacoom
Extremely rare tree, planted 1857, lot subdivided, tree murdered.					
Golden Weeping Willow	16'9"	72'	**89'**	295	Fife
Tree busted up, was removed.					
	15'3"	**78'**	67'	278	Avon
Tree removed, unknown reason.					
Ringleaf Willow	7'11"	**58'**	57'	167	Tacoma
Historic tree, removed for street repairs.					

*ERRATA FROM 1994 EDITION*_____

Mistakes in identification are always a problem when nearly 1,000 different kinds of trees are involved. While I have experts review these books before publication, the final editing decisions are mine, and I take full responsibility for the mistakes herein. There are dozens of name changes in the current register. Most of these have been cleared up by the new *Landscape Trees of North America* by Arthur Jacobson (1996) for the nomenclature. The major errors caught from 1994 are listed below.

The **Charlottae Crabapple** is actually the very similar **Klehm's Improved Bechtel Crabapple**.

The *Plumosa Albopicta* **Sawara Cypress** is a dwarf tree; the tree listed is *Plumosa Argentea*.

Pendula Vera is a **Lawson Cypress** with weeping twigs and branches; the tree listed previously was just a very weepy *Pendula*.

The giant **Greek Fir** at Point Defiance Park in Tacoma is actually the subspecies *graeca*, the **Apollo Fir**.

The tall **Crimean Pine** at the Wind River Arboretum near Carson is actually a **Corsican Pine**.

The large **Black Poplar** at Washington Park Arboretum in Seattle is actually a rare **Manchurian Poplar**.

The **Hollywood Plum** listed previously is unlike the true Hollywood, and thus appears to be unique; this time it is listed under the study name **"Seattle Hollywood"**.

The National Champion **Sitka Willow** is actually a **Scouler Willow**.

Whitebark Pine is a tree that lives near the upper limit of tree growth throughout the tall mountains of the Western states. Normally small, frequently multi-stemmed and broad-crowned, Whitebarks have a silhouette that offers a nice contrast to their frequent companion, Alpine Fir. Whitebark Pines are an important food source in timberline environments, their large seeds providing food for birds, bears, and small mammals. In Washington, Whitebarks are common only in the drier conditions found east of the main Cascade crest.

The tree pictured grows on the edge of a major serpentine formation in the Wenatchee Mountains. It is a giant, over five feet in diameter and 67 feet tall, sitting away from other trees in a meadow. Dave Braun is shown with the tree, which he discovered while hiking. Not far from this tree is a forest that contains a rich diversity of conifers. Alpine Larch, Alpine Fir, Mountain Hemlock, Engelmann Spruce, Douglas-fir, and Pacific Silver Fir all grow in this high-elevation forest. Within that grove is another Whitebark, even more impressive to me, as it is four feet in diameter, has no branches for 30 feet, and is 90 feet to the top - the tallest Whitebark Pine ever recorded!

Albin Dearing is seen here with the **Nolan Creek Redcedar.** Albin is the photographer for Davey Tree, which sponsors the National Big Tree Program through the American Forestry Association. Albin was on a trip to Washington in 1992 to photograph trees for the Big Tree Calendar, which features a different National Champion Tree for each month. The Nolan Creek tree is one of the easier subjects to photograph, since it is unobscured by a surrounding forest.

PLACE NAME INDEX

Entries in this index are alphabetized by the first word in the location heading for each tree. If you are visiting a new area, you can simply look up page numbers for any State Champion trees. Seattle and Tacoma Champions are too numerous to list. However, parks and cemeteries with more than a few State Champion trees are listed as sub-headings under cities. Similarly, trees at Washington Park Arboretum are too numerous to list in this index.

Aberdeen, 29, 32, 34, 58, 74
Acme, 63
Allen, 51, 82, 87
Alpine Lakes Wilderness, 8, 17
Anacortes, 10, 20, 57, 62, 99
Arlington, 2, 61, 64, 85, 96
Auburn, 20, 59, 64, 65, 95
Avon, 64, 81
Avondale, 66
Bainbridge Island, 73, 75, 81, 92
Bay Center, 6
Beacon Rock State Park, 1, 8, 17
Bellevue, 45, 46, 47, 48, 80, 81
Bellingham, 2, 26, 29, 48-49, 51, 53, 56-58, 68-69, 75, 77, 80, 82-83, 91, 99
Big Lake, 74, 84
Blaine, 2, 29
Bothell, 46
Bremerton, 36, 47, 51, 74, 91
Brinnon, 29, 38, 55, 68, 70-71, 90-91
Brush Prairie, 8, 97
Bryant, 15, 61
Burlington, 26, 46, 61
Camas, 37, 50, 77
Carnation, 6, 73, 96
Carson, 40, 45
Centralia, 33, 67, 81
Chehalis, 80
Chico, 34
Cicero, 40, 49
Clear Lake, 33
Clearview, 92
College Place, 33, 64, 69
Coupeville, 8

Dayton, 34, 58, 61, 66, 72
Duvall, 46
Edmonds, 41, 45, 66, 89
Elma, 57, 98
Enumclaw, 99
Everett, 27, 29, 34, 46-49, 73, 95
Fall City, 49, 74, 96
Federal Way, 72
Ferndale, 43, 57, 62, 96
Fife, 6, 98
Fir Island, 99
Fircrest, 34, 38, 81
Forks, 6, 8
Fort Lewis Military Reservation, 52
Fox Island, 45
Gifford Pinchot National Forest, 6, 8, 15, 17, 20
Gig Harbor, 20, 71
Glacier Peak Wilderness, 8, 10
Gold Bar, 2
Grandview, 98
Granite Falls, 34
Grays River, 28, 76
Greenbank, 49
Guemes Island, 15, 29
Hamilton, 13
Hoquiam, 25, 53
Issaquah, 41, 48, 82, 95
Juniper Dunes Wilderness, 11
Kent, 37, 41, 83, 98
Kirkland, 81, 90
Lake Forest Park, 26, 30, 74
Lake Stevens, 40, 92
Lakewood, 31, 35-38, 45, 57, 61, 63, 70, 74, 83, 88, 90

Laurel, 40
Leavenworth, 30
Longview, 27, 34, 43, 45, 47, 48, 51, 57, 67, 72, 76
Lowden, 79
Lyman, 73
Lynnwood, 30, 42, 43, 45, 57, 73, 74
Marysville, 25
Maury Island, 1, 10, 20
Mead, 15
Mercer Island, 51, 74
Milton, 34, 35, 81
Monroe, 31, 37, 48, 83
Montesano, 34, 35, 47, 96
Mt Adams Wilderness, 17
Mt Baker National Forest, 13
Mt Rainier National Park, 6
Mt St Helens National Volcanic Monument, 8
Mt Vernon, 6, 27, 34, 56, 58, 68, 76, 81, 83, 91, 94
Mountlake Terrace, 29, 79
Newhalem, 35, 36, 40
Newport, 1
Nisqually Wildlife Refuge, 6
North Bend, 76
Old Nisqually, 1
Olympia, 6, 26, 28, 34, 41, 45, 46, 48, 49, 75, 87, 88, 91, 93, 96, 99
Olympic National Forest, 1, 6, 8, 19
Olympic National Park, 6, 8, 10, 15, 17, 19
Orting, 25, 26, 29, 31
Parkland, 26, 32, 41, 97
Pe Ell, 6

QUICK REFERENCE
WHERE ARE WASHINGTON'S LARGEST TREES?

This tally shows the largest concentrations of Champion Trees. Due to the large number of introduced trees, cities are prominently represented. Within these cities, parks often have the highest concentrations of Champions. Many native Champs are located in Olympic National Park.

Cities			Parks	
1	Seattle	540	Washington Park Arboretum, Seattle	208
2	Tacoma	175	University of Washington, Seattle	48
3	Walla Walla	66	Point Definace Park, Tacoma	33
4	Puyallup	41	Wright Park, Tacoma	33
5	Spokane	32	Olympic National Park	23
6	Bellingham	22	Pioneer Park, Walla Walla	21
7	Olympia	19	Volunteer Park, Seattle	20
8	Vashon/Maury Island	18	USFS Arboretum, Wind River	18
9	Lakewood	17	Finch Arboretum, Spokane	16
10	Mt Vernon	17	Carl S English Gardens, Seattle	12

QUICK REFERENCE
WASHINGTON'S LARGEST INTRODUCED TREES

During the 1950s the American Forestry Association established a point system designed to compare trees. It has proved to be a useful system for comparing trees since it includes height, circumference and crown spread (see discussion on page xii and critiques on pages 4, 7 and 21). Most people use diameter because it can be seen and easily measured. These are the ten largest introduced tree species in Washington identified by these two methods.

	AFA Points			Circumference (feet and inches)	
1	Giant Sequoia	531		Giant Sequoia	32'6"
2	Lombardy Poplar	524		Lombardy Poplar	30'5"
3	Bolleana Poplar	386		Silver Maple	22'6"
4	Hybrid Plane	385		Monterey Cypress	22'4"
5	Silver Maple	375		Northern Catalpa *	22'0"
6	Northern Catalpa *	371		Bolleana Poplar	21'7"
7	American Chestnut *	366		Spanish Chestnut	21'5"
8	Monterey Cypress	363		American Chestnut *	20'7"
9	Corolina Poplar	357		Black Locust	20'3"
10	Black Locust	357		Hybrid Plane	20'2"

* National Champion

Port Angeles, 13
Port Townsend, 45, 97
Prairie, 57
Preston, 35
Pullman, 32
Puyallup, 26, 27, 29, 32, 34, 37, 40, 42, 44, 46-49, 51, 53, 58, 64, 66-67, 78, 81-83, 85-86, 92-93, 96-99
Quinault, 8
Redmond, 29, 31, 34, 49, 59, 74, 98
Renton, 26, 46, 47, 73
Richland, 59, 88
Richmond Beach, 29, 35, 54, 55, 67, 70, 72, 94
Ridgefield, 89
Rockport, 27, 38
Royal City, 87
Seattle, most pages
  Acacia Cemetery, 45, 51
  Calvary Cemetery, 31
  Carl S English Gardens, 32, 41, 60, 68, 74, 75, 78, 79, 85
  Cowen Park, 34, 93, 99
  Discovery Park, 41
  Evergreen Cemetery, 36, 48, 85, 89, 91
  Golden Gardens Park, 1, 10, 61, 63
  Green Lake Park, 20, 28, 46, 53, 56, 75, 77, 98
  Interlaken Park, 35, 53, 88
  Kinnear Park, 63, 64, 79
  Kubota Gardens Park, 62, 68, 83
  Leschi Park, 26, 48
  Lincoln Park, 20, 40, 41, 42, 59
  Me-Kwa-Mooks Park, 61, 69
  Mt Baker Park, 78
  Mt Pleasant Cemetery, 35, 73
  Ravenna Park, 52, 62, 81
  Seattle Pacific University, 2, 73,

78, 84
Seward Park, 6, 13, 20, 83, 88
University of Washington, 26-27, 31, 37-38, 40-45, 52, 54, 58, 59, 61-62, 75-77, 81, 83-85, 87, 93, 98
Volunteer Park, 27, 31, 38, 51, 54-57, 61, 62, 69, 72, 81, 83, 92
Washelli Cemetery, 37, 38, 70, 80
Washington Park Arboretum, most pages
Woodland Park Zoo, 31, 35, 38, 42, 74, 76
Sedro Woolley, 2, 26, 27, 33, 45, 48, 73, 76, 81, 82, 84
Selah, 31
Shelton, 46
Skagit City, 74, 95
Skagit Iasland, 10
Snohomish, 25, 30, 32, 34, 46, 49, 96
Spokane, 8, 10, 15, 34, 36, 41, 43, 44, 49, 50-55, 60-62, 68-70, 80, 81, 91, 98
Stanwood, 35, 83, 98
Steilacoom, 49, 57, 65, 68, 79, 87
Sultan, 38, 57, 63
Sumner, 33, 38, 46, 51, 52, 67, 76, 83, 85
Sylvana, 6, 13
Tacoma, most pages
  Calvary Cemetery, 49, 51
  Jefferson Park, 70, 90, 95
  Lincoln Park, 81
  Mountain View Cemetery, 38, 57, 61, 67, 83
  New Tacoma Cemetery, 61
  Old Tacoma Cemetery, 26, 57, 68, 70

Pt Defiance Park, 13, 26, 28, 33, 34, 38, 41, 45, 47, 50, 51, 53, 55, 60, 62, 67, 68, 70, 78, 80, 83, 84, 86, 89, 91, 92, 93, 98
Wapato Park, 1
Wright Park, 26-28, 30, 33, 41, 43, 46, 48-50, 57, 60, 63, 66-69, 73, 75-76, 81, 94, 97, 99
University of Puget Sound, 34, 66, 93
Toledo, 15, 49
Toppenish, 26, 56, 63, 83
Tukwila, 31, 35, 38, 52, 63, 99
Tumwater, 41
Umatilla National Forest, 1, 6, 13
Vancouver, 8, 34, 36, 38, 54, 74, 79, 81, 90, 94, 96, 97
Vashon Island, 15, 41, 45, 46, 54, 55, 57, 58, 74, 79, 81, 84
Walla Walla, 1, 2, 6, 8, 19-20, 26-29, 31-34, 37, 43, 49-50, 53, 56, 58-61, 64, 69, 74, 76, 78-79, 83, 85, 86, 88, 90, 93, 94, 95
Washougal, 25, 49, 58, 83
Wauna, 53, 82
Wenatchee, 29, 53, 61, 69, 83, 86
Wenatchee National Forest, 11, 13
Wind River, 51-53, 62, 80-83, 89, 91-92
Winslow, 20
Wishram, 26
Woodinville, 6, 26, 31, 37, 40, 46, 48, 64, 78, 90
Woodland, 1, 6, 15, 31, 33, 43, 46, 65, 67, 84, 90, 93, 95
Yakama Indian Resrevation, 15
Yakima, 29, 34, 42, 43, 51, 56, 77, 85, 97
Zillah, 11

GENERAL INDEX

This index is organized alphabetically by both common name and scientific name. Due to the splitting of native trees from introduced trees, and the different sections a tree might be in, finding a tree will often require the index. The main body of the book is organized alphabetically by common name, which can cause difficulty when only the scientific name is known. For example, the genus *Prunus* is found under many different headings: native cherries (page 6), Almonds (page 25), Apricot (page 26), Cherries (pages 35-40), Laurel (page 63), and Plums (pages 84-85). **Boldface** page numbers indicate a photo or drawing.

Abies, 8-9, 51-53
  alba, 51
  amabilis, 8 **9, 10**
  balsamea, 51
  x *bornmuelleriana*, 51
  bracteata, 52
  cephalonica var. *graeca*, 51 **52**
  cilicica, 51
  concolor, 53
    f. *violacea*, 51
  firma, 52
  fraseri, 51
  grandis, 8 **9, 24**
  holophylla, 52
  homolepis, 52
  x *insignis*, 51
  koreana, 52
  lasiocarpa, 8 **vi, 9**
  magnifica, 51
    var. *shastensis*, 53
  nebrodensis, 53
  nordmanniana, 51
  numidica, 51
  pindrow, 52 **52**
  pinsapo, 51, 52, 53
    'Glauca', 51
    ssp. *marocana*, 52
  procera, 8 **9, 10, 21**
    'Glauca Prostrata', 53
  veitchii, 53
Acer, 13, 15, 31, 67-73
  buergerianum, 71
  campestre, 68 **68**
  capillipes, 72
  cappadocicum, 67, 68 **68**
    ssp. *lobelii*, *68*
  circinatum, 15
  cissifolium, 71
  crataegifolium, 72
  davidii, 72
    'Hersii', 72
  x *freemanii* 'Armstrong', 67
  ginnala = *tartaricum* ssp. *ginnala*
  glabrum var. *douglasii*, 15
  griseum, 70
  hersii = *davidii* 'Hersii'
  japonicum 'Aconitifolium', 70
  lobelii = *A. cappadocicum* ssp. *lobelii*
  macrophyllum, 13, 68, 69 **12**
    'Kimballiae', 68
    'Seattle Sentinel', 69
  miyabei, 68
  mono = *pictum*
  monspessulanum, 68

  negundo, 31 **48**
    'Elegans', 31
    'Variegatum', 31
  nigrum = *A. saccharum* ssp. *nigrum*
  palmatum, 70-71
    ssp. *amoenum*, 71
    'Atropurpureum', 70 **71**
    'Bloodgood', 70
    'Burgundy Lace', 70
    var. *dissectum*
      f. *atropurpureum*, 71
    'Kagiri Nashiki', 70
    'Matsu Kaze', 70
    'Osakazuki', 70
    'Sango Kaku', 70
    'Tsuma Beni', 71
    'Whitney Red', 71
    'Wou Nishiki', 71
  pensylvanicum, 72
    'Erythrocladum', 72
  pictum, 70
  platanoides, 68, 69
    'Crimson King', 68
    'Erectum', 68
    'Faassen's Black', 68
    'Schwedleri', 69
  pseudoplatanus, 73
    'Atropurpureum', 73
    f. *variegatum*, 73
  rubrum, 67, 69
    'Columnare', 67
  rufinerve, 72
  saccharinum, 68, 69 **69**
    'Wieri Laciniatum', 68
  saccharum, 69, 73 **69**
    'Newton Sentry', 69
    ssp. *nigrum*, 67
    'Sweet Shadow', 73
  sempervirens, 68
  shirasawanum 'Aureum', 70
  sieboldianum, 71
  tataricum ssp. *ginnala*, 70
  tegmentosum, 72 **72**
  tetramerum var. *lobatum*, 72
  truncatum, 71
Aesculus, 32, 60-61
  californica, 32
  x *carnea*, 60, 61
    'Briotii', 60
  flava, 32
  glabra, 32
  hippocastanum, 60 **60**
    'Baumannii', 60
  indica, 60
  x *neglecta* 'Erythroblastos', 61
  octandra = *flava*

  pavia, 32
  turbinata, 60
AFA Points, x
Ailanthus altissima, 94 **94**
Alaska Cedar, see Cedar, Alaska
Albizia julibrissin, 90
Alder, 1, 25
  European, 25
  Italian, 25
  Imperial = Royal
  Mountain, 1
  Red, 1 **1**
  Royal, 25
  Sitka, 1
  White, 1
Almond, 25 **25**
  Double-flowered Hybrid, 25
  Hall's Hardy, 25
Alnus, 1, 25
  cordata, 25
  glutinosa, 25
    'Imperialis', 25
  rhombifolia, 1
  rubra, 1 **1**
  sinuata, 1
  tenuifolia, 1
Amelanchier, 17, 89
  alnifolia, 17 **2**
  laevis, 89
Amur Cork Tree, 79
Apple, 25
  Common, 25
  Gravenstein, 25
Apricot, 26
Aralia, Castor, 26
Araucaria araucana, 74 **73**
Arboretum Map, 102
Arborvitae, 26
  Columnar, 26
  Dark Golden = Golden Spike
  Douglas, 26
  Extra Gold = Golden Western
  George Peabody, 26
  Gold Column, 26
  Golden = George Peabody
  Golden Pyramid = Gold Column
  Golden Spike, 26
  Golden Western, 26
  Hiba, 26
  Hybrid, 26
  Japanese, 26
  Oriental, 26
  Pale Gold = Pale Gold Siberian
  Pale Gold Siberian, 26
  Variegated Hiba, 26
  Zebra, 26
Arbutus menziesii, 13 **iii, 12**

Ash, 1, 26-27
    Caucasian, 26
    European, 27
    Flame, 27
    Flowering, 27
    Green, 27
    Oregon, 1 **2**
    Weeping, 27
    White, 27 **27**
Asimina triloba, 79
Aspen, Quaking, 1
Azara microphylla, 27
Azara, Boxleaf, 27
Bald-Cypress, 28
    Pond, 28
Basswood, 28
    American, 28
    White, 28
Bayberry, Pacific, 2
Beech, 28-29
    American, 28 **27**
    Columnar, 28
    Copper, 29 **28**
    European, 29 **28**
    Fernleaf, 29
    Purple, 29
    Purple Oakleaf, 29
    Purple Tricolor, 29
    Weeping, 29
Betula, 2, 29-31
    albo-sinensis
        var. *septentrionalis*, 29
    alleghaniensis, 31 **30**
    lenta, 31
    lutea = alleghaniensis
    nigra, 31 **30**
    occidentalis, 2
    papyrifera, 2, 30
        var. *commutata*, 2 **1**
    pendula, 29, 30, 31
        'Crispa', 29 **29**
        'Fastigiata', 29
        'Laciniata' = 'Crispa'
        'Purpurea', 30
        'Youngii', 31 **29**
    platyphylla var. *japonica*, 30
    populifolia, 30
    pubescens, 29
    utilis var. *jacquemontii*, 30
Birch, 2, 29-31
    Chinese Paper, 29
    Columnar, 29
    Cutleaf Weeping, 29 **29**
    Downy, 29
    European White, 30
    Gray, 30
    Jacquemont, 30
    Japanese White, 30
    Paper, 30
    Purple, 30
    River, 31 **30**
    Sweet, 31
    Water, 2
    Western Paper, 2 **1**
    Yellow, 31 **30**
    Young's Weeping, 31 **29**

Box-elder, 31 **48**
    Variegated, 31
    Yellow Variegated, 31
Boxwood, 32
Buckeye, 32
    California, 32
    Ohio, 32
    Red, 32
    Yellow, 32
Butternut, 33
Buxus sempervirens, 32
Calocedrus decurrens, 34 **33**
Carpinus, 59-60
    betulus, 59, 60 **59**
        'Fastigiata', 60
    caroliniana, 59
    laxiflora, 59
    orientalis, 60
Carya, 57, 79
    glabra, 57
    illinoinensis, 79
    laciniosa, 57
    ovalis, 57
    ovata, 57
    tomentosa, 57 **58**
Cascara, 2
Castanea, 40-41
    x *blaringhemii*, 41
    dentata, 40 **40**
    mollissima, 41
    sativa, 41
Catalpa, 33
    bignonoides, 33
        'Nana', 33
    x *erubescens*, 33
    fargesii var. *duclouxii*, 33
    ovata 'Flavescens', 33
    speciosa, 33 **31**
Catalpa, 33
    Ducloux, 33
    Hybrid, 33
    Northern, 33 **31**
    Southern, 33
    Umbrella, 33
    Yellow, 33
Cedar, 4, 6, 34-35
    Alaska, 6 **3**
    Atlantic White, 34
    Atlas, 34 **31**
    Blue Atlas, 34 **32**
    Blue Sentinel, 34
    Cock's Comb Japanese, 34
    Deodar, 34 **32**
    Eastern Red Cedar, 34
    Golden Alaska, 34
    Golden Atlas, 34
    Golden Japanese, 34
    Golden Deodar, 34
    Incense, 34 **33**
    Japanese, 34 **33**
    Japanese Plume, 34
    Lebanese, 34
    Northern White, 34
    Port Orford, 34, 46
    Robust Deodar, 34
    Sentinel, 34

Variegated Alaska, 34
Variegated Atlantic White, 34
Weeping Alaska, 35
Weeping Blue Atlas, 35
Western Redcedar, 4, 6 **4, 5, 22, 101, 106**
Cedrus, 34-35
    atlantica, 34, 35 **31**
        'Aurea Robusta', 34
        f. *fastigiata*, 34
        'Glauca', 34 **32**
        'Glauca Fastigiata', 34
        'Glauca Pendula', 35
    deodara, 34 **32**
        'Aurea', 34
        'Robusta', 34
    libani, 34
Celtis, 8, 53
    occidentalis, 53
    reticulata, 8
Cercidiphyllum, 61
    japonicum, 61
    magnificum, 61
Cercis, 87
    canadensis, 87 **88**
        'Forest Pansy', 87
    siliquastrum, 87
Chamaecyparis, 6, 34-35, 45-48
    lawsoniana, 34, 46-47
        'Allumii', 46
        'Aureovariegata, 46
        'Columnaris', 46
        'Columnaris Glauca' =
            'Columnaris'
        'Elwoodii', 46
        'Erecta Glaucescens', 46
        'Erecta Viridis', 46 **47**
        'Fletcheri', 46
        'Fraseri', 46
        'Glauca', 46
        'Gracilis', 46
        'Hillieri', 46
        'Intertexta', 46
        'Lutea', 46
        'Lycopdioides', 47
        'Pendula', 47
        'Pendula Vera', 105
        'Pottenii', 47
        'Stewartii', 47
        'Tamariscifolia', 47
        'Triomf van Boskoop', 47
        'Versicolor', 47
        'Westermannii', 47
        'Wisselii', 47 **47**
        'Youngii', 47
    nootkatensis, 6, 34, 35 **3**
        'Lutea', 34
        'Pendula', 35
        'Variegata', 34
    obtusa, 45
        'Crippsii', 45
        'Gracilis', 45
    pisifera, 48
        'Aurea', 48
        'Boulevard', 48
        'Filifera', 48

'Filifera Aurea', 48
'Plumosa', 48 **48**
'Plumosa Albopicta', 105
'Plumosa Argentea', 48
'Plumosa Aurea', 48
'Squarrosa', 48
thyoides, 34
'Variegata', 34
Cherry, 6, 35-40, 45
Accolade, 37
Amanogawa, 38
Autumnalis Rosea, 37
Birchbark, 35
Bird, 35
Bitter, 6
Black, 35
Choke, 6, 35
Western, 6
Choshu-hizakura, 38
Common, 35
Cornelian-cherry, 45
Daybreak, 36 **35**
Double-Flowered Mazzard, 36
Fugenzo, 38
Fuji, 36
Goldbark, 36
Higan, 37
Hillier, 36
Hisakura, 38
Hokusai, 38 **36, 39**
Horinji, 38
Japanese Flowering, 37-39
Japanese Hill, 36
Korean Hill, 36
Kwanzan, 38 **39**
Mikuruma-gaeshi, 38
Miyama, 36
Mt Fuji, 38
Naden, 36
Ojochin, 38
Oshima, 36
Pandora, 37
Pendula, 37
Pedula Plena Rosea, 37
Pendula Rubra, 37
Pink Perfection, 38
Pink Weeping, 37
Sargent, 40
Sato Zakura, 38-39
Shirofugen, 38
Shirotae/Mt. Fuji, 38
Shogetsu, 38 **39**
Spaeth, 40
St Lucie, 40
Stellata, 37
Tai Haku, 38 **39**
Taoyoma, 38
Temari, 38
Ukon, 38 **39**
Weeping Higan, 37
Western Choke, 6
Whitcomb, 37 **37**
Winter Flowering, 37
Yoshino, 40 **35**
Chestnut, 40-41
American, 40 **40**

Chinese, 41
European = Spanish
Spanish, 41
Sweet American, 41
China-fir, 41
Blue, 41
Chinquapin, Golden, 6
Chokecherry, 6, 35
Western, 6
Chrysolepis chrysophylla, 6
Circumference, measuring, vii
Cladrastis, 98
kentukea, 98
lutea = kentukea
platycarpa, 98
Clereodendrum trichotomum, 53
Cornus, 6, 45, 49
alternifolia 'Argentea', 49
controversa, 49
'Eddie's White Wonder', 45
florida, 45. 49
f. *rubra*, 49
kousa, 45
mas, 45
nuttallii, 6
occidentalis, 6
Corylus, 10, 55
avellana, 55
'Contorta', 55
'Tortuosa' = 'Contorta'
cornuta var. *californica*, 10
colurna, 55
Cotinus, 90
coggygria, 90
'Foliis Purpureis', 90
obovatus, 90
Cottonwood, 6, 86
Black, 6 **23**
Eastern, 86
Crabapple, 6, 41-44
Almey, 43
Aspiration, 41
Bechtel, 41
Blanche Ames, 41 **42**
Charlottae, 105
Cherry, 41
Chinese Flowering, 41
Cutleaf, 41
Dawson, 41 **42**
Dolgo, 41
Dorothea, 41
Echtermeyer, 43
Eleyi, 43
Hillier, 41
Himalayan, 41
Hopa, 43 **43**
Hupeh, 41
Japanese Flowering, 41
Kaido, 42
Katheriine, 42
Klehm's Improved Bechtel, 42
Limoinei, 43
Liset, 43
Manchurian, 42
Oregon, 6
Parkman, 42

Pink Beauty, 43 **43**
Pillar, 42
Prairie, 42
Prince Georges, 42
Profusion, 43
Purple, 43
Radiant, 43
Red Jade, 42
Redvein, 43
Rosy-bloom, 43
Royalty, 43
Sargent, 42
Scheidecker, 42
Siberian, 44
Siberian Crab Hybrid, 44
Sweet, 44
Toringo, 44
Van Eseltine, 44
Yellow Autumn, 44
Zumi, 44
Crape-Myrtle, 44
Crataegus, 8, 10, 54-55, 56
'Autumn Glory', 54
columbiana, 10
crus-galli, 54
douglasii, 8
laevigata, 56
'Bicolor', 56
'Crimson Cloud', 56
'Masekii', 56
'Paul's Scarlet', 56 **56**
'Plena', 56 **56**
'Punicea', 56
'Rosea Flore Pleno', 56
x *lavallei*, 54
mollis, 54
monogyna, 56
f. *pendula*, 56
x *mordenensis* 'Toba', 54
nitida, 54
phaenopyrum, 55
punctata, 54, 55
'Aurea', 55
sanguinea, 54
viridis 'Winter King', 55
Crown Spread, measuring, x
Cryptomeria japonica, 34 **33**
'Cristata', 34
'Elegans', 34
'Sekkan-Sugi', 34
Cunninghamia lanceolata, 41
'Glauca', 41
Cupressus, 45
arizonica, 45 **44**
var. *glabra*, 45
glabra = arizonica var. *glabra*
goveniana var. *pygmaea*, 45
macrocarpa, 45 **44**
nevadensis, 45
sempervirens, 45
'Stricta', 45
X *Cupressocyparis leylandii*, 45
'Castlewellan', 45
'Contorta', 45
'Leighton Green', 45
'Silver Dust', 45

Cydonia oblonga, 87
Cypress, 45-48
    Arizona, 45 **44**
    Arizona Smooth, 45
    Castlewellan, 45
    Columnar Italian, 45
    Contorted Leyland, 45
    Golden Hinoki, 45
    Hinoki, 45
    Italian, 45
    Lawson, 46-47
        Allumii, 46
        Aureovariegata, 46
        Columnaris, 46
        Columnaris Glauca =
            Columnaris
        Ellwoodii, 46
        Erecta Glaucescens, 46
        Erecta Viridis, 46 **47**
        Fletcheri, 46
        Fraseri, 46
        Glauca, 46
        Gracilis, 46
        Hillieri, 46
        Intertexta, 46
        Lutea, 46
        Lycopodioides, 47
        Pendula, 47
        Pendula Vera, 105
        Pottenii, 47
        Stewartii, 47
        Tamariscifolia, 47
        Triomf van Boskoop, 47
        Versicolor, 47
        Westermannii, 47
        Wisselii, 47 **47**
        Youngii, 47
    Leyland, 45
    Mendocino, 45
    Monterey, 45 **44**
    Piute, 45
    Slender Hinoki, 45
    Sawara, 48
        Aurea, 48
        Boulevard, 48
        Filifera, 48
        Filifera Aurea, 48
        Plumosa, 48 **48**
        Plumosa Albopicta, 105
        Plumosa Argentea, 48
        Plumosa Aurea, 48
        Squarrosa, 48
    Smooth Arizona = Arizona Smooth
    Variegated Leyland, 45
Date-Plum, 79
Davidia involucrata, 49
Diospyros, 79
    kaki, 79
    lotus, 79
    virginiana, 79
Dogwood, 6, 45, 49
    Cornelian-cherry, 45
    Eastern, 45
    Eddie's White Wonder, 45
    Kousa, 45
    Pacific, 6

Pink, 49
Silver Pagoda, 49
Table, 49
Western, 6
Douglas-fir, 6-7, 49
    Bigcone, 49
    Coast, 6-7 **ii, xii, 7**
    Dwarf, 49
    Rocky Mountain, 6
    Weeping, 49
Dove Tree, 49
Elaeagnus angustifolia, 88
Elder, 8, 49
    Blue, 8
    European Black, 49
    Pacific Red, 8
Elm, 49-50
    American, 49 **50**
    Camperdown, 49
    Chinese, 49
    Columnar American, 49
    Cornish, 49
    English, 49
    European White, 49 **50**
    Guernsey, 49
    Huntingdon, 49
    Jersey = Guernsey
    Rock, 49
    Scotch = Wych
    Siberian, 50
    Smooth-leaved, 50
    Tornado, 50
    Weeping American, 50
    Wych, 50
Embothrium coccineum
    var. *lanceolatum*, 53
Empress Tree, 50
    Lilac, 50
Epaulette Tree, 51
    Smooth-barked, 51
Eriobotrya japonica, 65 **64**
Errata from Previous Edition, 105
Eucalyptus, 103
    gunnii, 103
    perriniana, 103
Eucalyptus, 103
    Cider Gum, 103
    Spinning Gum, 103
Eucommia ulmoides, 54
Eucryphia x *nymanensis*
    'Nymansay', 51
Eucryphia, Nymans Hybrid, 51
Euodia, 51
    hupehensis = *Tetradium daniellii*
    velutina = *Tetradium daniellii*
Euodia, 51
    Hupeh, 51
    Szechwan, 51
Euonymous, 90
    europaeus, 90
    hamiltonianus
        ssp. *sieboldianus*, 90
Fagus, 28-29
    grandifolia, 28 **27**
    sylvatica, 28, 29 **28**
        'Asplenifolia', 29

    'Dawyck', 28
    'Pendula', 29
    f. *purpurea*, 29 **28**
    'Purpurea Tricolor', 29
    'Rohanii', 29
    'Roseomarginata' = 'Purpurea
        Tricolor'
Ficus carica, 51
Fig, Edible, 51
Fir, 8-9, 51-53
    Algerian, 51
    Alpine, 8 **vi, 9**
    Apollo, 51 **52**
    Balsam, 51
    Blue Colorado, 51
    Blue Spanish, 51
    Bornmueller, 51
    California Red, 51
    Caucasian, 51
    Cilician, 51
    European Silver, 51
    Fraser, 51
    Grand, 8 **9, 24**
    Greek, 105
    Hybrid Caucasian, 51
    Korean, 52
    Manchurian, 52
    Momi, 52
    Moroccan, 52
    Nikko, 52
    Noble, 8 **9, 10, 21**
    Pacific Silver, 8 **9, 10**
    Pindrow, 52 **52**
    Santa Lucia, 52
    Shasta Red, 53
    Sicilian, 53
    Spanish, 53
    Spreading Noble, 53
    Veitch, 53
    White, 53
Fire-tree, Chilean, 53
Former Record Trees, 103
Fraxinus, 1, 26-27
    americana, 27 **27**
    angustifolia, 26
        'Flame', 27
    excelsior, 27
        'Pendula', 27
    latifolia, 1 **2**
    ornus, 27
    oxycarpa = *angustifolia*
    pennsylvanica, 27
Ginkgo biloba, 53
Ginkgo, 53
Gleditsia triacanthos, 58-59
    'Elegantissima', 58
    f. *inermis*, 59 **59**
    'Ruby Lace', 58
    'Sunburst', 58
Glorybower, Harlequin, 53
Golden-chain, 53
    Alpine, 53
    Chimaeric, 53
    Common, 53
    Hybrid, 53
Golden Raintree, 53

Gymnocladus dioicus, 61
Hackberry, 8, 53
   Common, 53
   Netleaf, 8
Halesia carolina var. *monticola*, 90
Hardy Rubber-tree, 54
Hawthorn, 8, 10, 54-56
   Autumn Glory, 54
   Bicolor, 56
   Black, 8
   Cockspur, 54
   Columbia, 10
   Common, 56
   Crimson Cloud, 56
   Dotted, 54
   Downy, 54
   English, 56
   Lavalle, 54
   Masekii, 56
   Midland, 56
   Oneseed, 56
   Paul's Scarlet, 56 **56**
   Pendula, 56
   Pink, 56
   Plena, 56 **56**
   Punicea, 56
   Rosea Flore Pleno, 56
   Shining, 54
   Siberian, 54
   Toba, 54
   Washington, 55
   Weeping, 56
   Winter King, 55
   Yellow Fruited, 55
Hazel, 10, 55
   California, 10
   Corkscrew, 55
   European, 55
   Turkish, 55
   Twisted = Corkscrew
Hemlock, 10, 11 55, 57
   Carolina, 55 **58**
   Eastern, 55
   Jenkin's, 55
   Mountain, 10, 11 **11**
   Northern Japanese, 55
   Sargent's Weeping, 57 **55**
   Southern Japanese, 57
   Western, 10, 11 **i, 11**
Height, estimating, ix
Height, measuring, viii
Hickory, 57
   Mockernut, 57 **58**
   Pignut, 57
   Red, 57
   Shagbark, 57
   Shellbark, 57
Holly, 57-58
   American, 57
   Balearic, 57
   Camellia Leaved Highclere, 57
   English, 57
   Golden King Highclere, 57
   Golden Queen, 57
   Gold Variegated, 57
   Hedgehog, 57
   Hodgins, 57

   Japanese, 57
   Orange Berried, 57
   Perry's Silver Weeping, 58
   Red Twigged, 58
   Silver Hedgehog, 58
   Silver Variegated, 58
   Variegated Hedgehog = Silver
      Hedgehog
   Yellow Berried, 58
Honeylocust, 58-59
   Globe, 58
   Ruby Lace, 58
   Sunburst, 58
   Thornless, 59 **59**
Hop-hornbeam, 59
   Eastern, 59
   European, 59
Hornbeam, 59-60
   American, 59
   European, 59 **59**
   Looseflower, 59
   Oriental, 60
   Pyramidal, 60
Horsechestnut, 60-61
   Briot, 60
   Common, 60 **60**
   Double Flowered, 60
   Indian, 60
   Japanese, 60
   Red, 61
   Sunrise, 61
How to Measure a Tree, vii
Ilex, 57-58
   × *altaclerensis*, 57
      'Balearica', 57
      'Camelliaefolia', 57
      'Golden King', 57
      'Hodginsii', 57
   aquifolium, 57, 58
      'Amber', 57
      'Argentea Marginata', 58
      'Argentea Marginata
         Pendula', 58
      'Aurea Marginata', 57
      f. *bacciflava*, 58
      'Ferox', 57
      'Ferox Argentea', 58
      'Golden Queen', 57
      'Rubricaulis Aurea', 58
   crenata, 57
   opaca, 57
Ironwood, Persian, 61
Judas-tree, 87
Juglans, 33, 95-97
   ailanthifolia, 96
      var. *cordiformis*, 96
   cathayensis, 95
   cinerea, 33
   hindsii, 97
   hindsii x *nigra*, 97
   x *intermedia*, 96
   major, 95
   mandshurica, 96
   microcarpa, 97
   nigra, 95 **100**
   'Paradox', 97 **96**
   regia, 96 **95**

Juniper, 10-11, 34, 61
   Blue Eastern, 61
   Chinese, 61
   Eastern Red Cedar, 34
   Farges', 61
   Hollywood, 61 **60**
   Irish, 61
   Keteleer, 61
   Meyer, 61
   Pyramidal Chinese, 61
   Rocky Mountain, 10
   Variegated Chinese = Variegated
      Hollywood
   Variegated Hollywood, 61
   Weeping Eastern, 61
   Western, 11
   Young's Golden, 61
Juniperus, 10-11, 34, 61
   chinensis, 61
      'Aurea', 61
      'Kaizuka' = var. *torulosa*
      'Keteleerii', 61
      'Pyramidalis', 61
      var. *torulosa*, 61 **60**
         'Variegata', 61
      'Variegata' = var. *torulosa*
         'Variegata'
   communis 'Hibernica', 61
   occidentalis, 11
   scopulorum, 10
   squamata
      var. *fargesii*, 61
      'Meyeri', 61
   virginiana, 34, 61
      f. *glauca*, 61
      'Pendula', 61
Kalopanax, 26
   pictus = *septemlobus*
   septemlobus, 26
Katsura, 61
   Magnificent, 61
Kentucky Coffeetree, 61
Koelreuteria paniculata, 53 **54**
+ *Laburnocytisus adamii*, 53
Laburnum, 53
   alpinum, 53
   anagyroides, 53
   × *watereri*, 53
Lagerstroemia indica, 44
Larch, 11, 13, 62
   Alpine, 11
   Dunkeld, 62
   European, 62 **62**
   Golden, 62
   Japanese, 62
   Polish, 62
   Tamarack, 62
   Weeping European, 62
   Western, 13 **23**
Largest Trees
   Introduced, 108
   Locations, 107
   Native, 21

Larix, 11, 13, 62
    decidua, 62 **62**
        'Pendula', 62
        ssp.. *polonica*, 62
    x *eurolepis*, 62
    kaempferi, 62
    laricina, 62
    lyallii, 11
    occidentalis, 13 **23**
Laurel, 62-63
    Bay, 62
    English, 63
    Magnolia-leaved, 63
    Portugal, 63 **62**
Laurus nobilis, 62
Ligustrum lucidum, 87
Lilac, Japanese Tree, 63
Linden, 63-64
    Bigleaf, 63
    Crimean, 63
    European, 63
    Littleleaf, 64 **63**
    Silver, 64
    Silver Pendent, 64
    Spectacular, 64
Liquidambar, 93
    orientalis, 93
    styraciflua, 93
        'Variegata', 93
Liriodendron tulipifera, 94 **94**
Lithocarpus densiflorus, 94
    f. *attenuato-dentatus*, 94
Locust, 64-65
    Black, 64 **64**
    Columnar, 64
    Contorted, 64
    Globe, 64
    Golden, 65
    Hybrid Mexican, 65
    Idaho, 65
    Pink, 65
    Singleleaf, 65
Loquat, 65 **64**
Maackia, 65
    amurensis var. *buergeri*, 65
    chinensis, 65
Maackia, 65
    Chinese, 65
    Japanese, 65
Maclura pomifera, 78
Madrona, Pacific, 13 **iii**, **12**
Magnolia, 65-67
    acuminata, 65, 67
        var. *subcordata*, 67
    campbellii, 65
    cylindrica, 65
    dawsoniana, 65
    denudata, 67
    fraseri, 65, 66
        ssp. *pyramidata*, 66
    grandiflora, 65, 66, 67
        'Goliath', 66
        'Victoria', 67
    hypoleuca, 66
    kobus, 66
        var. *loebneri* 'Merrill', 66

macrophylla, 65
x *proctoriana* 'Wada's Memory', 67
salicifolia, 67
    'Wada's Memory' = *M.* ·
    x *proctoriana* 'Wada's Memory'
sargentiana var. *robusta*, 66
x *soulangiana*, 65, 66, 67
    f. *alba*, 67
    'Alexandrina', 65
    'Brozzonii', 65
    'Rustica Rubra', 66
sprengeri, 67
stellata 'Rosea', 66
x *thompsoniana*, 67
tripetala, 67
x *veitchii*, 67 **66**
virginiana, 67
x *wieseneri*, 67
Magnolia, 65-67
    Alexandrina, 65
    Anhwei, 65
    Bigleaf, 65
    Brozzoni, 65
    Campbell, 65
    Cucumber Tree, 65
    Dawson, 65
    Evergreen, 65
    Fraser, 65
    Goliath, 66
    Japanese Silverleaf, 66
    Kobus, 66
    Merrill, 66
    Pink Star, 66
    Pyramid, 66
    Rustica Rubra, 66
    Sargent, 66
    Saucer, 67
    Sprenger, 67
    Sweet Bay, 67
    Thompson, 67
    Umbrella, 67
    Veitch, 67 **66**
    Victoria, 67
    Wada's Memory, 67
    Watson, 67
    White Saucer, 67
    Willowleaf, 67
    Yellow Cucumber Tree, 67
    Yulan, 67
Malus, 6, 25, 41-44
    x *adstringens*, 44
    'Almey', 43
    baccata, 41, 42, 44
        'Aspiration', 41
        var. *himalaica*, 41
        var. *mandshurica*, 42
    'Blanche Ames', 41 **42**
    coronaria, 42, 44
        'Charlottae', 103
        'Klehm's Improved Bechtel, 42
    x *dawsoniana*, 41 **42**
    'Dolgo', 41
    x *domestica*, 25
        'Gravenstein', 25
    'Dorothea', 41
    'Echtermeyer', 43

'Eleyi', 43
x *floribunda*, 41
fusca, 6
halliana 'Parkmanii', 42
'Hillieri' = *M.* x *scheideckeri* 'Hillieri'
'Hopa', 43 **43**
hupehensis, 41
ioensis, 41, 42
    'Plena', 41
'Katherine', 42
'Lemoinei', 43
'Liset', 43
x *micromalus*, 42
niedzwetzkyana = *M. sieversii*
    'Niedzwetzkyana'
'Pink Beauty', 43 **43**
'Prince Georges', 42
'Profusion', 43
x *purpurea*, 43
'Radiant', 43
'Red Jade', 42
x *robusta*, 41
'Royalty', 43
sargentii, 42
x *scheideckeri*, 41, 42
    'Hillieri', 41
sieboldii var. *arborescens*, 44
sieversii 'Niedzwetzkyana' 43
spectabilis 'Plena', 41
x *sublobata*, 44
toringoides, 41
tschonoskii, 42
'Van Esseltine', 44
x *zumi*, 44
Maple, 13, 15, 67-73
    Amur, 70
    Armstrong, 67
    Bigleaf, 13 **12**
    Black, 67
    Bloodgood, 70
    Burgundy Lace, 70
    Chinese Stripebark, 72
    Coliseum, 67 **68**
    Columnar Red, 67
    Coral Bark, 70
    Cretan, 68
    Crimson King, 68
    Cutleaf Bigleaf, 68
    Cutleaf Silver, 68
    Douglas, 15
    East Asian, 70
    English Field, 68 **68**
    Erect Norway = Mt Hope
    Faassen's Black, 68
    Fern Leaf, 70
    Golden Fullmoon, 70
    Gray-budded Snakebark, 72
    Hawthorn Leaved Snakebark, 72
    Hers's Stripebark, 72
    Japanese, 70-71
    Japanese Red, 70 **71**
    Lobel's, 68
    Manchurian Stripebark, 72
    Matsu Kaze, 70
    Miyabe, 68
    Montpelier, 68

Moosewood, 72
Newton Sentry, 69
Norway, 69
Osakazuki, 70
Painted, 70
Paperbark, 70
Père David's Stripebark, 72
Pink Variegated Japanese, 70
Red, 69
Red Lace Leaf, 71
Red Nail Japanese, 71
Red Snakebark, 72
Red-Twigged Moosewood, 72
Schwedler, 69
Seattle Sentinel, 69
Seven Lobed Japanese, 71
Shantung, 71
Siebold, 71
Silver, 69 **69**
Snakebark, 72
Stripebark, 72
Sugar, 69 **69**
Sweet Shadow, 73
Sycamore, 73
Trident, 71
Variegated Sycamore, 73
Vine, 15
Vineleaf, 71
Wineleaf Sycamore, 73
Whitney Red, 71
Wou Nishiki, 71
Mayten, 73
Maytenus boaria, 73
Medlar, 73
Mespilus germanica, 73
Metasequoia glyptostroboides, 88 **89**
Monkey-puzzle, 74 **73**
Morus, 74
    alba, 74
        'Pendula', 74
    nigra, 74
Mountain-ash, 15, 74
    Cascade, 15
    Chinese, 74
    Columnar, 74
    European, 74 **73**
    Hupeh, 74
    Japanese, 74
    Pratt, 74
    Sitka, 15
    Weeping European, 74
Mulberry, 74
    Black, 74
    Weeping, 74
    White, 74
Myrica californica, 2
Myrtle, 44, 74
    Crape-myrtle, 44
    Oregon, 74
Nominating a Tree, x
Nomination Form, 120
Northofagus, 90
    antarctica, 90
    dombeyi, 90
    obliqua, 90
Nyssa sylvatica, 95

Oak, 15, 74-78
    Armenian, 74
    Bamboo-leaf, 74
    Black, 76
    Bur, 76
    California Black, 74
    Canyon Live, 74
    Cherrybark, 76
    Chestnut, 76
    Chinese Cork, 74
    Chinquapin, 76
    Coast Live, 75
    Cork, 75
    Cypress, 75
    Daimyo, 75
    Durmast, 75
    Eastern North American, 76-77
    English, 75 **75**
    Holm, 75
    Huckleberry, 75
    Interior Live, 75
    Laurel, 76
    Northern Pin, 76
    Oregon White, 15 **13**
    Pin, 76 **75**
    Red, 76 **77**
    Sawtooth, 75
    Scarlet, 76 **101**
    Shingle, 76 **77**
    Shumard, 77
    Silverleaf, 78
    Southern Red, 77
    Swamp Chestnut, 77
    Swamp White, 77
    Turkish, 78
    Ubame, 78
    Valley, 78
    Water, 77
    White, 77
    Willow, 77
Osage-Orange, 78
Osmanthus, 78
    x *fortunei*, 78
    heterophyllus, 78
Osmanthus, 78
    Hybrid, 78
    Japanese, 78
Ostrya, 59
    carpinifolia, 59
    virginiana, 59
Oxydendrum arboreum, 90
Pagoda Tree, Japanese, 79
Palm, Chinese Windmill, 79 **78**
Parrotia persica, 61
Paulownia tomentosa, 50
    'Lilacina', 50
Pawpaw, 79
Pear, 79
    Asian, 79
    Callery, 79
    Common, 79 **78**
    Hybrid Snow, 79
    Willow Leaved Hybrid, 79
Pecan, 79

Persimmon, 79
    American, 79
    Date-plum, 79
    Japanese, 79
Phellodendron amurense, 79
Phellodendron, 79
    Amur Cork Tree, 79
Phillyrea latifolia, 87
Photinia, 79-80
    fraseri, 79, 80
        'Birmingham', 79
    serratifolia, 80
        'Nova Lineata', 80
Photinia, 79-80
    Birmingham, 79
    Chinese, 80
    Fraser, 80
    Variegated, 80
Picea, 15, 17, 19, 91-93
    abies, 91, 92, 93
        'Cupressina', 91
        'Intermedia', 92
        'Inversa', 93 **92**
        f. *pyramidata*, 91
        f. *virgata*, 92 **92**
    asperata, 91
        var. *retroflexa*, 92
    brachytyla, 92
    breweriana, 91
    engelmannii, 17 **17**
    glauca, 93
    glehnii, 92
    jezoensis, 91, 93
        var. *hondoensis*, 91
    mariana, 91
    omorika, 92
    orientalis, 92 **91**
    polita, 92
    pungens, 91
        'Columnaris', 91
        f. *glauca*, 91 **91**
    rubens, 92
    sitchensis, 15, 18, 19 **xi, 16, 18**
    smithiana, 93
Pine, 14, 15, 17, 80-83
    Austrian, 80 **119**
    Bishop, 80
    Bristlecone, 80
    Chinese, 80
    Chinese White, 80
    Columnar Scots, 81
    Columnar White, 81
    Corsican, 81 **80**
    Coulter, 81
    Crimean, 81
    Digger, 81 **80**
    Eastern White, 81
    Himalayan White, 81
    Italian Stone, 81
    Jack, 81
    Japanese Black, 81
    Japanese Red, 81
    Japanese White, 81
    Jeffrey, 81 **82**
    Knobcone, 81
    Korean, 82

Lacebark, 82
Limber, 82
Loblolly, 82
Lodgepole, 15
Macedonian, 82
Maritime, 82 **82**
Mexican Piñon, 82
Monterey, 82
Montezuma, 83
Mountain, 83
Mugo, 83
Pitch, 83
Ponderosa, 14, 15 **14**
Red, 83
Scots, 83
Shore, 15
Shortleaf, 83
Southwestern White, 83
Sugar, 83
Swiss Stone, 83
Table Mountain, 83
Tanyosho, 83
Twisted White, 83
Virginia, 83
Weeping White, 83
Western White, 17 **13**
Whitebark, 17 **105**
Pinus, 14, 15, 17, 80-83
  albicaulis, 17 **105**
  aristata, 80
  armandii, 80
  attenuata, 81
  banksiana, 81
  bungeana, 82
  cembra, 83
  cembroides, 82
  contorta, 15
      var. *contorta*, 15
      var. *latifolia*, 15
  coulteri, 81
  densiflora, 81, 83
      'Umbraculifera', 83
  echinata, 83
  flexilis, 82
  jeffreyi, 81 **82**
  koraiensis, 82
  lambertiana, 83
  montezumae, 83
  monticola, 17 **13**
  mugo, 83
  muricata, 80
  nigra, 80, 81 **119**
      var. *corsicana*, 81 **80**
      ssp. *laricio* = var. *corsicana*
      ssp *pallasiana* = var *pallasiana*
      var. *pallasiana*, 81
  parviflora, 81
      'Glauca', 81
  peuce, 82
  pinaster, 82 **82**
  pinea, 81
  ponderosa, 14, 15 **14**
  pungens, 83
  radiata, 82
  resinosa, 83
  rigida, 83

  sabiniana, 81 **80**
  strobiformis, 83
  strobus, 81, 83
      'Fastigiata', 81
      'Pendula', 83
      'Torulosa', 83
  sylvestris, 81, 83
      'Fastigiata', 81
  taeda, 82
  tabulaeformis, 80
  thunbergii, 81
  uncinata, 83
  virginiana, 83
  wallichiana, 81
Pistacia chinensis, 83
Pistachio, Chinese, 83
Place Name Index, 107
Plane, 83-84
  Hybrid, 83
  Oriental, 84
  Pyramidal Hybrid, 84 **84**
Platanus, 83-84, 93
  x *acerifolia*, 83, 84
      'Pyramidalis', 84 **84**
  occidentalis, 93
  orientalis, 84
Plum, 84, 85
  Burbank Vesuvius, 85
  Cherry, 84
  Cistena, 85
  Hollywood, 105
  Japanese, 84
  Moser, 85 **85**
  Newport, 85
  Pissard, 85
  Purple, 85
  Purple Leaved, 85
  Seattle Hollywood, 85
  Shiro, 84
  Thundercloud, 85
  Trailblazer, 85
Poplar, 84, 86, 87
  Berlin, 84
  Black, 84, 105
  Bolleana White, 84
  Carolina, 86
  Cathay, 87
  Certines, 87
  Columnar Simon, 87
  Eugenei, 86
  Ghost, 87
  Gray, 87
  Japanese, 87
  Hybrid Black, 86
  Lombardy, 86, 87 **84**
  Manchurian, 87
  Marilandica, 86
  Regenerata, 86 **86**
  Robusta, 86
  Serotina, 86
  Serotina Aurea, 86
  Szechuan, 87
  Weeping Simon, 87
  White, 87
Populus, 1, 6, 84, 86, 87
  alba, 84, 87

  'Nivea', 87
  'Pyramidalis', 84
  x *berolinensis*, 84, 87
      'Certinensis', 87
  x *canadensis*, 86
      'Eugenei', 86
      'Marilandica', 86
      'Regenerata', 86 **86**
      'Robusta', 86
      'Serotina', 86
      'Serotina Aurea', 86
  x *canescens*, 87
  cathayana, 87
  'Certinensis' = *P.* x *berolinensis*
      'Certinensis'
  deltoides, 86
  maximowiczii, 87
  nigra, 84, 87
      'Afghanica', 87
      'Italica', 86, 87 **84**
      Thevestina' = 'Afghanica'
  simonii, 87
      'Fastigiata', 87
      'Pendula', 87
  songarica, 87
  szechuanica, 87
  tremuloides, 1
  trichocarpa, 6 **23**
Privet, 87
  Chinese, 87
  Mock, 87
Prunus, 6, 25, 26, 35-40, 63, 84, 85
  'Accolade', 37
  'Amanogawa', 38
  x *amygdalo-persica* =
      x *persicoides*
  armeniaca, 26
  avium, 35, 36
      'Plena', 36
  x *blireiana* 'Moseri', 85 **85**
  'Burbank Vesuvius' = 'Vesuvius'
  cerasifera, 84, 85
      'Hollywood', 105
      'Pissardii', 85
      'Thundercloud', 85
  'Choshu-hizakura', 38
  'Cistena', 85
  dulcis, 25 **25**
  emarginata, 6
  'Fugenzo', 38
  'Hillieri', 36
  'Hisakura', 38
  'Hokusai', 38 **36, 39**
  'Horinji', 38
  incisa, 36, 37
  jamasakura, 36, 38
  'Kwanzan', 38 **39**
  lannesiana, 38
  laurocerasus, 63
      'Magnoliaefolia', 63
  lusitanica, 63 **62**
  maackii, 36
  mahaleb, 40
  maximowiczii, 36
  'Mikuruma-gaeshi', 38
  'Mt Fuji', 38

'Newport', 85
'Ojochin', 38
padus, 35, 40
    'Spaethii', 40
'Pandora', 37
pendula, 37
    var. ascendens, 37
    'Pendula Plena Rosea', 37
    'Pendula Rubra', 37
    'Stellata', 37
x persicoides, 25
    'Hall's Hardy', 25
    'Roseoplena, 25
'Pink Perfection', 38
salicina, 84
    'Shiro', 84
sargentii, 40
'Seattle Hollywood', 85
serotina, 35
serrula, 35
serrulata, 38
    'Amanogawa', 38
    'Choshu-hizakura', 38
    'Fugenzo', 38
    'Hokusai', 38 **36, 39**
    'Horinji', 38
    'Kanzan', 38 **39**
    'Mikuruma-gaeshi', 38
    'Mt Fuji', 38
    'Ojochin', 38
    'Pink Perfection', 38
    'Shirofugen', 38
    'Shirotae', 38
    'Shogetsu', 38 **39**
    'Tai Haku', 38 **39**
    'Taoyoma', 38
    'Temari', 38
    'Ukon', 38 **39**
'Shirofugen', 38
'Shirotae', 38
'Shogetsu', 38 **39**
x sieboldii, 36
speciosa, 36, 38
x subhirtella, 37
    'Accolade' = P. 'Accolade'
    'Autumnalis Rosea', 37
    'Pandora' = P. 'Pandora'
    'Pendula' = P. pendula
    'Pendula Plena Rosea' =
        P. pendula 'Pendula Plena
        Rosea'
    'Pendula Rubra' = P. pendula
        'Pendula Rubra'
    'Stellata' = P. pendula 'Stellata'
    'Whitcomb', 37 **37**
'Tai Haku', 38 **39**
'Taoyoma', 38
'Temari', 38
'Trailblazer', 85
'Ukon', 38 **39**
verecunda, 36, 38
'Vesuvius', 85
virginiana, 6, 35
    var. demissa, 6
x yedoensis, 36, 40 **35**
    'Akebono', 36 **35**

Pseudocydonia sinensis, 87
Pseudolarix amabilis, 62
Pseudotsuga, 6, 7, 49
    macrocarpa, 49
    menziesii, 6, 7, 49
        var. glauca, 6
        var. menziesii, 6, 7 **ii, xii, 7**
        'Pendula', 49
        'Slavinii', 49
Pterocarya, 98
    fraxinifolia, 98
    stenoptera, 98
Pterostyrax, 51
    corymbosa, 51
    hispida, 51
Pyrus, 79
    calleryana, 79
    x communis, 79 **78**
    pyrifolia var. culta, 79
    salicifolia hybrid, 79
    x salvifolia, 79
Quercus, 15, 74-78
    acutissima, 75
    agrifolia, 75
    alba, 77
    bicolor, 77
    cerris, 78
    chrysolepis, 74
    coccinea, 76 **101**
    dentata, 75
    ellipsoidalis, 76
    falcata, 76, 77
        var. pagodifolia, 76
    garryana, 15 **13**
    hypoleucoides, 78
    ilex, 75
    imbricaria, 76 **77**
    kelloggii, 74
    laurifolia, 76
    lobata, 78
    macrocarpa, 76
    michauxii, 77
    muehlenbergii, 76
    myrsinifolia, 74
    nigra, 77
    palustris, 76 **75**
    petraea, 75
    phellos, 77
    phillyraeoides, 78
    pontica, 74
    prinus, 76
    robur, 75 **75**
        'Fastigiata', 75
    rubra, 76 **77**
    shumardii, 77
    suber, 75
    vaccinifolia, 75
    variabilis, 74
    velutina, 76
    wislizenii, 75
Quince, 87
    Chinese, 87
    Common, 87
Redbud, 87
    Eastern, 87 **88**
    Forest Pansy, 87
    Judas-Tree, 87

Redcedar, Western,
    see Cedar, Western redcedar
Red Cedar, Eastern,
    see Cedar, Eastern Red
Redwood, 88
    Blue, 88
    Cantab, 88
    Coast, 88 **88**
    Dawn, 88 **89**
Register of Washington State Big Trees,
    Introduced, 25
    Native, 1
Rehder Tree, 88
Rehderodendron macrocarpum, 88
Rhamnus purshiana, 2
Rhododendron macrophyllum, 17
Rhododendron, Pacific, 17
Rhus, 19, 93
    glabra, 19
    pottaninii, 93
    typhina, 93
Robinia, 64-65
    x ambigua, 65
        'Decaisneana', 65
        'Idahoensis', 65
    x holdtii, 65
    'Idaho' = R. x ambigua 'Idahoensis'
    pseudoacacia, 64, 65 **64**
        'Frisia', 65
        'Pyramidalis', 64
        'Tortuosa', 64
        'Umbraculifera', 64
        'Unifoliola', 65
Russian Olive, 88
Salix, 19-20, 98
    alba, 98
        'Argentea' = var. sericea
        var. calva = var. coerulea
        var. coerulea, 98
        var. sericea, 98
        var. vitellina, 98
    amygdaloides, 20
    babylonica, 98
        'Annularis', 98 **99**
        'Crispa' = 'Annularis'
    bebbiana, 19
    cinerea, 98
    daphnoides, 98
    elaeagnos ssp. angustifolia, 98
    hookeriana, 19
    lasiandra, 20
    lasiolepis, 19
    matsudana, 98
        'Tortuosa', 98
    pentandra, 98
    piperi, 20
    x rubens, 98
    scouleriana, 20 **19**
    x sepulcralis, 98
        'Chrysocoma', 98
        'Salamonii', 98
    x sericans, 98
    sitchensis, 20
Sambucus, 8, 49
    caerulea, 8
    callicarpa, 8
    nigra, 49

Sassafras albidum, 88
Sassafras, 88
Sciadopitys verticillata, 95
Sequoia, 89
    Blue, 89
    Giant, 89 **89**
    Weeping, 89
Sequoia sempervirens, 88 **88**
       'Cantab', 88
       f. *glauca*, 88
Sequoiadendron giganteum, 89 **89**
       'Glaucum', 89
       'Pendulum', 89
Serviceberry, 17, 89
    Allegheny, 89
    Western, 17 **2**
Service Tree, 89
    Devon, 89
    of Fountainbleau, 89
    True, 89
    Wild, 89
Silk Tree, 90
Silverbell, Mountain, 90
Smoke Tree, 90
    American, 90
    European, 90
    Purple, 90
Snowbell, 90
    Bigleaf, 90
    Japanese, 90
Sophora japonica, 79
       'Pendula', 79
Sorbus, 15, 74, 89, 97
    aria, 97
    aucuparia, 74 **73**
       f. *fastigiata*, 74
       'Pendula', 74
    commixta, 74
    devoniensis, 89
    domestica, 89
    hupehensis, 74
    intermedia, 97 **97**
    x *latifolia*, 89
    prattii, 74
    scalaris, 74
    scopulina, 15
    sitchensis, 15
    x *thuringiaca*, 97
    torminalis, 89
    'Wilfrid Fox', 97
Sorrel Tree, 90
Southern Beech, 90
    Antarctic, 90
    Dombey, 90
    Roble, 90
Spindle Tree, 90
    European, 90
    Japanese, 90
Spruce, 15, 17, 18, 19, 91-93
    Black, 91
    Blue, 91 **91**
    Brewer, 91
    Columnar Blue, 91
    Cypress Norway, 91
    Dragon, 91
    Engelmann, 17 **17**

Hondo, 91
Norway, 91
Oriental, 92 **91**
Pyramidal Norway, 92
Red, 92
Sakhalin, 92
Sargent, 92
Serbian, 92
Sitka, 1, 15, 18, 19 **xi, 16, 18**
Snakebranch, 92 **92**
Tapo Shan, 92
Tigertail, 92
Twisted Branch Norway, 92
Weeping Norway, 93 **92**
West Himalayan, 93
White, 93
Yeddo, 93
Stewartia, 93
    monadelpha, 93
    ovata, 93
    pseudocamellia, 93
Stewartia, 93
    Common, 93
    Mountain, 93
    Tall, 93
Styrax, 90
    japonicus, 90
    obassia, 90
Sumac, 19, 93
    Pottanin, 93
    Smooth, 19
    Staghorn, 93
Sweetgum, 93
    American, 93
    Golden, 93
    Oriental, 93
Sycamore, American, 93
Syringa reticulata, 63
Tall Trees, 23, 101
    Native, 23
    Exotic, 101
Tamarisk, 93
    Spring Flowering, 93
    Summer Flowering, 93
Tamarix, 93
    chinensis 'Plumosa', 93
    parviflora, 93
Tanoak, 94
    Cutleaf, 94
Taxodium, 28
    ascendens, 28
    distichum, 28
Taxus, 20, 99
    baccata, 99
       'Adpressa', 99
       'Aurea', 99
       'Fastigiata', 99
       'Fastigiata Aurea', 99
       'Repandans', 99
    brevifolia, 20 **20**
    x *media*, 99
Tetradium daniellii, 51
Thuja, 6, 26, 34
    occidentalis, 26, 34
       'Aureospicata', 26
       'Fastigiata', 26

       'Pyramidalis' = 'Fastigiata'
       'Douglasii Pyramidalis', 26
       'Lutea', 26
       'Wareana Lutescens', 26
    occidentalis x *plicata*, 26
    orientalis, 26
       'Elegantissima', 26
    plicata, 4, 6, 26 **4, 5, 22, 101, 106**
       'Aurea', 26
       'Zebrina', 26
       'Zebrina Extra Gold' = 'Aurea'
    standishii, 26
Thujopsis dolabrata, 26
    'Variegata', 26
Tilia, 28, 63-64
    americana, 28
    cordata, 64 **63**
    'Euchlora', 63
    x *euchlora* = 'Euchlora'
    x *europaea*, 63
    heterophylla, 28
    platyphyllos, 63
    x *spectabilis*, 64
    tomentosa, 64
       'Pendula', 64
       'Petiolaris' = 'Pendula'
Torreya nucifera, 94
Torreya, Japanese, 94
Trachycarpus fortunei, 79 **78**
Tree Measurements, vii
Tree of Heaven, 94 **94**
Tsuga, 10, 11, 55, 57
    caroliniana, 55 **58**
    canadensis, 55, 57
       'Jenkinsii', 55
       'Pendula', 57 **55**
    diversifolia, 55
    heterophylla, 10, 11 **i, 11**
    mertensiana, 10, 11 **11**
    sieboldii, 57
Tuliptree, 94 **94**
Tupelo, Black, 95
Ulmus, 49-50
    americana, 49, 50 **50**
       'Fastigiata', 49
       f. *pendula*, 50
       'Pendula' = f. *pendula*
    glabra, 49, 50
       'Camperdownii', 49
    x *hollandica* 'Vegeta', 49
    laevis, 49 **50**
    minor, 49, 50
       var. *cornubiensis*, 49
       'Gracilis', 50
       var. *sarniensis*, 49
       var. *vulgaris*, 49
    parvifolia, 49
    procera = *U. minor* var. *vulgaris*
    pumila, 50
    thomasi, 49
Umbellularia californica, 74
Umbrella-pine, Japanese, 95
Volume Measurements, ix
Walnut, 95-97
    Arizona, 95
    Black, 95 **100**

Cathay, 95
English, 96 **95**
Heartnut, 96
Hybrid English, 96
Japanese, 96
Manchurian, 96
Northern California Black, 97
Paradox, 97 **96**
Royal Hybrid, 97
Texas Black, 97
Whitebeam, 97
Finnish, 97
Swedish, 97 **97**
Wilfrid Fox, 97
Willow, 19-20, 98
Arroyo, 19
Bay, 98
Bebb, 19

Chinese, 98
Chinese Weeping, 98
Corkscrew, 98
Golden, 98
Golden Weeping, 98
Gray, 98
Hooker, 19
Hybrid Pussy, 98
Hybrid Weeping, 98
Hybrid White, 98
Narrow Leaved Rosemary, 98
Pacific, 20
Peachleaf, 20
Piper, 20
Ringleaf, 98 **99**
Scouler, 20 **19**
Silver, 98
Sitka, 20

Violet, 98
White, 98
Wingnut, 98
Caucasian, 98
Chinese, 98
Yellowwood, 98
Japanese, 98
Yew, 20, 99
English, 99
Golden English, 99
Golden Irish, 99
Hybrid, 99
Irish, 99
Shortleaf, 99
Spreading English, 99
Pacific, 20 **20**
Zelkova serrata, 99
Zelkova, 99

The state's largest **Austrian Pine** was this tree on the Whitman College campus in Walla Walla. The reign was short-lived, however, as a building was built on the tree's roots. The building extended to within a few feet of the trunk and the tree promptly died. Even a strong, healthy tree as this is susceptible to disturbance.

*BIG TREE NOMINATION FORM*_____

COMMON NAME OF TREE _____

SCIENTIFIC NAME OF TREE _____

CIRCUMFERENCE OF TRUNK (in feet and inches) _____

TREE HEIGHT (feet) _____

AVERAGE CROWN SPREAD (feet) _____

CONDITION OF TREE _____

EXACT LOCATION (photocopy of map/street atlas helpful) _____
_____
_____
_____

NAME AND ADDRESS OF NOMINATOR _____
_____
_____
_____

NAME AND ADDRESS OF OWNER _____
_____
_____
_____

DATE AND METHODS OF MEASUREMENT _____
_____
_____
_____

PHOTOGRAPH ENCLOSED ____YES ____NO

SEND NOMINATIONS TO:

> Washington State Big Tree Program
> Box 352100
> Anderson Hall
> University of Washington
> Seattle, WA 98195

WASHINGTON BIG TREE PROGRAM